D1044983

Fourier methods in crystallography

Wiley monographs in crystallography

Fourier Methods
in Crystallography

G. N. Ramachandran

Director, Centre of Advanced Study in Biophysics
University of Madras, India, and
Professor of Biophysics, University of Chicago

R. Srinivasan

Professor of Crystallography
Centre of Advanced Study in Biophysics
University of Madras, India

Wiley - Interscience

a division of John Wiley & Sons, Inc.

New York · London · Sydney · Toronto

To the leader of Modern Indian Science, Professor C. V. Raman, who initiated the senior author into the mysteries of diffraction theory and Fourier transforms.

Preface

The subject of x-ray crystallography is essentially concerned with the diffraction of x-rays. It is well known that the phenomenon of diffraction is intimately connected with the subject of Fourier transformation. In fact, the distribution of scattering matter in real space is connected with the distribution of intensity in various directions in the diffracted beam by an operation of Fourier transformation. Since a Fourier transform has, as its inverse operation, another Fourier transform, the problem of determining the distribution of matter from the distribution of intensity is a process involving such a transformation. It is natural, therefore, that the subjects of Fourier series and Fourier transforms play a fundamental part in calculating the x-ray diffraction patterns and in the inverse problem of determining the molecular structure from the diffraction pattern.

The present monograph deals essentially with aspects of these problems, particularly for a periodic crystal. Some of the basic formulas of Fourier transformation, such as the idea of convolution of two functions and of its Fourier transform, are considered in the first chapter. In view of the intimate relationship between a structure and its Fourier transform, the latter can in fact be taken to represent the structure. Thus one can consider structures represented by various functions of the Fourier transform of a given structure. This method of approach leads to interesting possibilities of getting back the structure from the observed x-ray intensities. As is well known, a simple Fourier inversion of the intensity pattern gives only the self-convolution of the structure with its inverse, but not the structure itself. The problem of deriving the structure from this inversion can be approached either in the actual crystal space or in the space of the Fourier transform. The former procedure is best carried out in terms of the theory of images, an approach so ably considered by Professor Buerger in his book, *Vector Sets*. The present monograph is in a sense complementary to his, in that it develops procedures for deriving the structure from the Patterson function via Fourier methods, applied straight to x-ray intensities.

Our laboratory has been deeply interested in this problem for more than a decade. In a sense this volume is a connected account of the methods developed in Madras, with a discussion of related topics and extensions of these methods, but it is mainly theoretical and outlines only the methods. There has been no attempt to indicate how the Fourier series are computed or how the structure is refined from observed data by Fourier methods. On the other hand, the methods of isomorphous replacement and anomalous dispersion are discussed in detail, with special reference to their application in the solution of phase problem.

We are very grateful to our students and our colleagues, with whom we have been working over the last ten years and more at the University of Madras, for the stimulation that we have received in these studies. In particular, we should like to thank Dr. R. Chandrasekaran and Dr. A. R. Kalyanaraman for their help in checking the manuscript. The generous assistance of the authorities of the University of Madras who made the facilities at the Centre available to us is gratefully acknowledged. Both of us had the benefit of spending a period in the United States in 1968, which gave us time to collect our thoughts and put them down in this book. One of us (G.N.R.) would like to thank the chairman and members of the Biophysics Department of the University of Chicago for the facilities afforded there. The other (R.S.) is grateful to Professor Henry Koffler, chairman, and Professor M. G. Rossmann of the Department of Biological Sciences, Purdue University, for the facilities provided during January to September, 1968. We should also like to express our appreciation of the advice and assistance given by Professor Martin J. Buerger in the preparation and editing of the monograph. We are grateful also to the National Institutes of Health, U.S.A., University Grants Commission, India, and the Jawaharlal Nehru Memorial Fund, India, for financial assistance. We should particularly like to thank Mrs. Bernice P. Manaster and Mr. K. P. Ramachandran for secretarial assistance, Mr. V. R. Sambandam for the drawings and Dr. C. Ramakrishnan and Mr. T. Srikrishnan for assistance in proof reading.

Madras, India   G. N. Ramachandran
February, 1970   R. Srinivasan

Contents

Contents

Contents

Fourier methods in crystallography

1

Fourier representation of electron density

Introduction

One of the principal methods of investigating the ultrastructure of matter is to perform experiments on the scattering of waves. The most notable example is the scattering of x-rays by matter, which has been widely employed for the elucidation of the structure of molecules and crystals at the atomic level. The theory of the interrelation between the scattering phenomena and the structure of the scattering material is most simply formulated by making use of the principle of Fourier transformation. This subject has probably reached its greatest height in its development with regard to its application to the diffraction of x-rays by matter—in particular by solids in crystalline form. We shall briefly outline the Fourier-transform theory of diffraction in this chapter with special reference to the diffraction by crystals.

Theory of diffraction by an assemblage of particles and a continuous medium

Consider a monochromatic parallel beam of x-rays and let the point P represent a scattering center, which may be either an atom or an electron (Fig. 1). The problem is to calculate the resultant scattered radiation in a given direction making an angle of 2θ with the incident direction (Fig. 1a). Let $\mathbf{s}_i$ be the unit vector along the direction of the incident radiation and $\mathbf{s}_r$ be a unit vector along the direction of the scattered radiation. The phase of the wave scattered by the point P in the direction $\mathbf{s}_r$ is ahead of the phase of

1

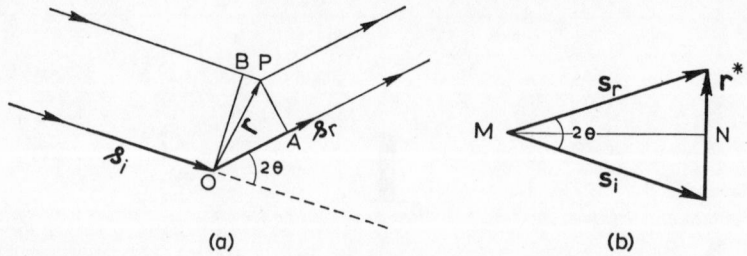

Fig. 1. (a) Relation between incident and scattered waves. s_i and s_r are unit vectors along incident and scattered directions. (b) The incident-wave vector S_i and the scattered-wave vector S_r are of magnitude $1/\lambda$. S_r may be considered to be obtained by specular reflection of S_i in the plane MN.

the wave scattered by a point at the origin O, by a quantity δ, given by

$$\delta = \frac{2\pi}{\lambda}(AO - PB) = \frac{2\pi}{\lambda}\mathbf{r} \cdot (\mathbf{s}_r - \mathbf{s}_i), \tag{1}$$

where $\mathbf{r}$ is the position vector of P with O as origin. Defining the wave vectors S_i and S_r by

$$\mathbf{S}_i = \frac{1}{\lambda}\mathbf{s}_i, \qquad \mathbf{S}_r = \frac{1}{\lambda}\mathbf{s}_r, \tag{2}$$

the phase difference, δ, may be written as

$$\delta = 2\pi\mathbf{r} \cdot (\mathbf{S}_r - \mathbf{S}_i) = 2\pi\mathbf{r} \cdot \mathbf{r}^*, \tag{3a}$$

where

$$\mathbf{r}^* = \mathbf{S}_r - \mathbf{S}_i. \tag{3b}$$

Thus, if a is the amplitude of the wave scattered by P when the amplitude of the incident wave is unity, then the wave from P may be written, both in amplitude and phase, as

$$a \exp(2\pi i\mathbf{r}^* \cdot \mathbf{r}),$$

where the phase of the wave scattered from the origin is taken to be zero. The quantity a may be called the scattering factor of P and is in general a function of $\mathbf{r}^*$, namely $a(\mathbf{r}^*)$.

Set of N scattering points

Suppose there is a set of N scattering points at positions $\mathbf{r}_j$ ($j = 1$ to N) of scattering factors a_j. Then the total wave scattered is obtained by summing the waves scattered by all the scatterers in the given direction. Also, the scattering phenomenon (namely the relationship between the directions of the incident and the scattered waves, and the wavelength of the radiation) is represented by the vector **r***. Hence the total scattered amplitude corresponding to **r*** is

$$F(\mathbf{r}^*) = \sum_{j=1}^{N} a_j \exp(2\pi i \mathbf{r}^* \cdot \mathbf{r}_j). \tag{4}$$

Continuous medium

In the case of a continuous medium let $\rho(\mathbf{r})$ denote the scattering density at a point **r**. Then the amplitude of the wave scattered by the matter in a volume element dv at **r** is $\rho(\mathbf{r})dv$ and the wave scattered by dv is given in amplitude and phase by $\rho(\mathbf{r})dv \exp(2\pi i \mathbf{r}^* \cdot \mathbf{r})$. Therefore the total scattering in the scattering direction is given by

$$F(\mathbf{r}^*) = \int \rho(\mathbf{r}) \exp(2\pi i \mathbf{r}^* \cdot \mathbf{r}) \, dv_r, \tag{5}$$

where the subscript r for the volume element denotes the fact that the volume element is measured in real space, viz., the space in which **r** is defined.

Thus the wave scattered by a continuous medium can be represented as the Fourier transform of the scattering density, and it also follows, from the theory of Fourier transforms, that the scattering density is, in turn, given by the inverse Fourier transform, namely

$$\rho(\mathbf{r}) = \int F(\mathbf{r}^*) \exp(-2\pi i \mathbf{r}^* \cdot \mathbf{r}) \, dv_{r^*}, \tag{6}$$

where **r*** and dv_{r^*} are measured in the Fourier (or reciprocal) space (see Appendix A for a proof of this). We shall presently see the significance of the term " reciprocal " for this space.

It will be useful to know the magnitude of the vector **r***, which is defined by $\mathbf{r}^* = \mathbf{S}_r - \mathbf{S}_i$. Let the scattering angle, i.e., the angle between $\mathbf{S}_i$ and $\mathbf{S}_r$, be denoted by 2θ. Then from the Fig. 1b it follows that

$$|\mathbf{r}^*| = 2|\mathbf{S}_r| \sin \theta = \frac{2 \sin \theta}{\lambda}. \tag{7}$$

It will thus be seen that the dimensions of the length of the vector **r*** are in cm^{-1}, if λ is measured in centimetres (or Å^{-1}, if λ is measured in angstroms). Otherwise, **r*** is a vector exactly like **r**, and can have any magnitude or direction. Hence, we say that the vector **r*** defines a reciprocal space. The scalar product of **r** and **r*** is a dimensionless number.

The geometry of Fig. 1b is interesting. The scattering vector **r*** makes equal angles with the directions of the incident and scattered waves. If a plane is considered to be drawn normal to **r*** (shown by MN in Fig. 1b), then the scattered wave may be considered to have been obtained by specular reflection in this plane.

We shall now give a brief summary of the application of the Fourier-transform approach to the study of the diffraction of x-rays by a crystal. As a preliminary to this, the formulas regarding the scattering by an atom will also be given to make the treatment complete.

Scattering by atoms

Scattering by a free electron

The electrons in the atoms and molecules are mainly responsible for the scattering of x-rays by matter. A single free electron is effectively a point

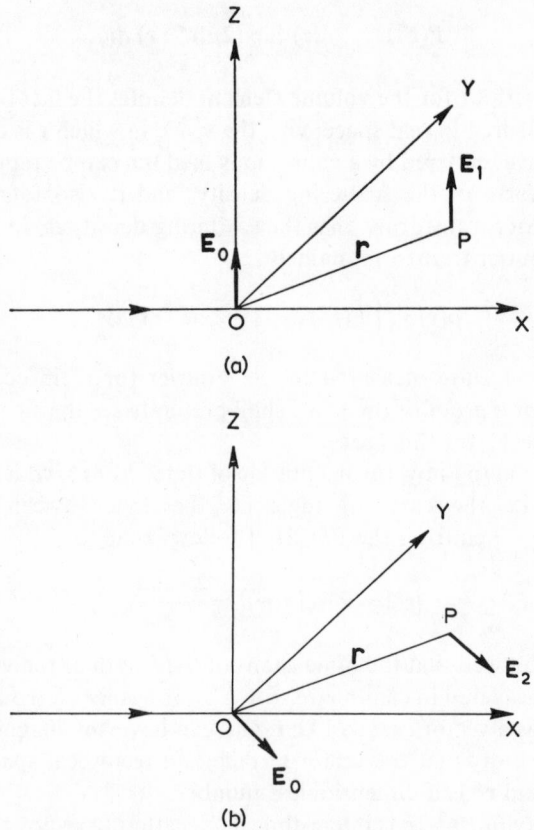

Fig. 2. Coordinate axes for the description of the scattered amplitude at the point P by a free electron at the origin O. (a) Electric vector parallel to OZ; (b) in the plane XY.

scatterer for x-rays, since its dimensions are small ($\sim 10^{-13}$ cm) compared to the wavelength of the x-rays ($\sim 10^{-8}$ cm) used in the usual diffraction studies. Consider an electron situated in the path of a parallel beam of x-rays traveling in the direction OX (Fig. 2). Classically, the electron, when accelerated by the electric vector $\mathbf{E}_0$ of the incident wave, reradiates the electromagnetic radiation with the same wavelength. If the electric vector $\mathbf{E}_0$ is along the direction OZ, the scattered amplitude at a point P at a distance $\mathbf{r}$ in the XY plane is given[8] by

$$\mathbf{E}_1 = -\frac{e^2}{mc^2r}\mathbf{E}_0, \tag{8}$$

where e and m are the electronic charge and mass, respectively, and c is the velocity of light. The corresponding intensity is given by

$$I_1 = k^2 I_0, \tag{9}$$

where

$$k = \frac{e^2}{mc^2r} \tag{10a}$$

and

$$I_0 = E_0^2. \tag{10b}$$

If the incident wave has the electric vector parallel to the XY plane (i.e., in the direction OY) then the intensity is given by

$$I_2 = k^2 I_0 \cos^2 2\theta, \tag{11}$$

where 2θ is the scattering angle, i.e., the angle between the incident and scattered directions.

For a beam of unpolarized radiation the intensity is obviously given by

$$I = k^2 \left(\frac{1 + \cos^2 2\theta}{2}\right) I_0. \tag{12}$$

This is the classical expression known as the Thomson formula for the scattering by a free electron. It is most convenient to express the intensity of x-rays scattered by any given system in terms of the scattering by a single free electron as the unit. The scattering of any sample thus expressed would be a pure number which stands for the ratio of the scattered intensity by the given system to that which will be scattered by a single free electron under the same conditions. It is also well to remember[8] that the phase of the scattered wave for the free electron is ahead of the phase of the incident wave by an angle π.

Scattering by the electron distribution in an atom

According to well-known wave-mechanical theory, the distribution of electronic matter in an atom is continuous. Consequently, the total amplitude of the wave scattered by an atom, $f(\mathbf{r}^*)$, can be expressed in terms of the electron-density distribution $\rho(\mathbf{r})$ of the electron cloud. Since the amplitude scattered by a single free electron is taken as the unit, the amplitude of the wave scattered by $\rho(\mathbf{r})\,dv$ is itself equal to $\rho(\mathbf{r})\,dv$. Then according to the previous section we have

$$f(\mathbf{r}^*) = \int \rho(\mathbf{r}) \exp(2\pi i \mathbf{r} \cdot \mathbf{r}^*)\, dv_r. \tag{13}$$

The quantity $f(\mathbf{r}^*)$ is called the atomic scattering factor. Its value, for different values of $\mathbf{r}^*$, can be obtained by integrating the right-hand side of (13) over the volume of the atom (in principle up to infinity, but in practice over a region at the boundary of which the electron density is negligibly small).

In general, the electron density of a free atom can be assumed to be spherically symmetrical and the integral (13) then reduces to the form of (14), in which it is a function only of the magnitude of $\mathbf{r}^*$, namely r^*, which is equal to $2(\sin\theta)/\lambda$ by (7). Thus, as has been shown in the Appendix F (A35),

$$f(\mathbf{r}^*) = \int 4\pi r^2 \rho(r) \frac{\sin(2\pi r r^*)}{2\pi r r^*}\, dr. \tag{14}$$

The radial distribution function $4\pi r^2 \rho(r)$ has been obtained from wave-mechanical theory for various atoms and the values of $f[(\sin\theta)/\lambda]$ calculated from these are available in tabular form.[2,10]

The coherent-scattering amplitude derived above is referred to as the normal scattering amplitude and is valid strictly under two conditions, namely, (a) that the electron density distribution in an atom is spherically symmetrical, and (b) the scattering by each element of volume dv is $\rho\,dv$ times that of the free electron. Both these conditions may not be fully satisfied in practice under certain circumstances, and appropriate modifications are then necessary. In most cases, however, the changes involved may turn out to be small and can be neglected unless a very accurate analysis is aimed at.

For instance, condition (a) is violated[4–6,11–14] when some of the electron shells are incomplete. For example, in the case of the $2p$ orbital, which has an axis of symmetry, it becomes necessary to consider two cases, one in which the vector $\mathbf{r}^*$ is parallel to the above direction and another in which it is perpendicular to it. The contribution to the integral (14) by such an electronic distribution is therefore different from that of a completely spherically symmetric distribution. A closely related situation that affects the spherical symmetry of density distribution is the bonding between atoms, which necessarily exists in most molecules, an effect which cannot be neglected for accurate work.[4-6]

The second assumption (*b*) necessarily implies that the element of charge $\rho \, dv$ may be taken to scatter in phase with a free electron at the same location. On the other hand, we know that the electrons in atoms are not free, but are bound to the nucleus. There are thus distinct characteristic frequencies for the atom, and phenomena connected with resonance absorption occur if the incident x-ray wavelength is close to such an absorption edge. In fact, detailed theoretical treatment shows (e.g., see James,[8] p. 135, et seq.) that the atomic scattering factor is not independent of the frequency of the incident wave. The normal atomic scattering factor is only an approximation, valid for frequencies far removed from any of the absorption edges of the atom. For the general case, the atomic scattering factor f is given by the expression

$$f = f^0 + \delta f' + i \, \Delta f'', \tag{15}$$

where $\delta f'$ and $\Delta f''$ are frequency-dependent correction terms. We shall discuss these in detail in Ch. 11. The fact that there is an imaginary component $i \, \Delta f''$ in (15) implies that there exists a component with a phase shift of $\pi/2$. Thus while the total real component $f' = (f^0 + \delta f')$ represents a component in phase with the scattering by a free electron, there exists an " out of phase " component $\Delta f''$ which is $\pi/2$ ahead of f'. The contributions $\delta f'$ and $\Delta f''$ arise basically from the innermost electrons, which are relatively tightly bound to the nucleus, for elements of medium atomic number and for x-ray wavelengths of the order of 1 Å.

The occurrence of $\delta f'$ and $\Delta f''$ and their consequences are generally referred to as anomalous-dispersion effects. In fact, anomalous dispersion has become a powerful tool in x-ray analysis in recent years.

Fourier transforms

From what has been discussed earlier, it follows that if $\rho(\mathbf{r})$ is the electron-density distribution in a piece of matter, the scattered x-ray wave in a direction defined by the scattering vector $\mathbf{r}^*$ is given by

$$F(\mathbf{r}^*) = \int \rho(\mathbf{r}) \exp (2\pi i \mathbf{r}^* \cdot \mathbf{r}) \, dv_r \tag{16}$$

where $|F|$ is measured taking the scattering amplitude of a single free electron under the same conditions to be equal to unity. The right-hand side of (16) is a general Fourier transformation integral, and $F(\mathbf{r}^*)$ is called the Fourier transform of $\rho(\mathbf{r})$.

As is clear from the foregoing sections, it is the Fourier transform of a structure that is directly related to the properties of the scattered wave rather than the structure itself. In fact, the relation (16) is a basic one applicable to all the scattering processes where one is interested in the coherent scattered wave produced by a given system. Thus, for instance, the above relation is true for optical, neutron, and electron diffraction also, with appropriate scaling constants. The mathematical aspects of such transformations have been well studied (see for example Titchmarsh[19]). We shall consider here some of the basic properties of Fourier transforms.

We may first start with the relation (16) as the definition of the Fourier transform $F(\mathbf{r}^*)$ of $\rho(\mathbf{r})$. It is obtained by multiplying $\rho(\mathbf{r})$ by the kernel $\exp(2\pi i \mathbf{r} \cdot \mathbf{r}^*)$ and integrating over the entire volume in the space $\mathbf{r}$ in which $\rho(\mathbf{r})$ is defined. Symbolically we may write

$$F(r^*) = T[\rho(\mathbf{r})], \qquad (17)$$

where the operator symbol T denotes multiplication by $\exp(2\pi i \mathbf{r} \cdot \mathbf{r}^*)$ and integration over the appropriate volume of space. It can be shown that $\rho(\mathbf{r})$ is also expressible in terms of $F(\mathbf{r}^*)$ in an inverse relation of a very similar type. This inverse relation is

$$\rho(\mathbf{r}) = \int F(\mathbf{r}^*) \exp(-2\pi i \mathbf{r} \cdot \mathbf{r}^*) \, dv_{r^*}, \qquad (18)$$

and it corresponds to the inverse operation represented symbolically by

$$\rho(\mathbf{r}) = T^{-1}[F(\mathbf{r}^*)]. \qquad (19)$$

The proof of this is given in Appendix A. It may be emphasized that the inverse transformation corresponds to multiplication by the quantity $\exp(-2\pi i \mathbf{r} \cdot \mathbf{r}^*)$ and integration over the appropriate space (in this case that of the vector $\mathbf{r}^*$). In relations (16) and (18) we could as well have interchanged the signs in the exponents and the two relations together would still have been valid. What is important to remember, however, is that the pair of inverse transformations involve the use of opposite signs; e.g., as defined above, the forward transformation T is associated with a $+$ sign in $\exp(+2\pi i \mathbf{r}^* \cdot \mathbf{r})$ and the inverse transformation is associated with a $-$ sign in $\exp(-2\pi i \mathbf{r}^* \cdot \mathbf{r})$. In many of the standard books on Fourier transforms (e.g., Titchmarsh[19]) the factor 2π does not occur in the exponent, but it occurs in a natural way in all diffraction problems, and in fact, when it is included, some of the integral relations involving the Fourier transforms take simple symmetrical forms.

In our present study, we shall use the Fourier transform relations as defined by our equations (16) and (18). The more important of the formulas required for our study, using this form, are either proved as required, or are stated in Appendix A.

As already mentioned, the "spaces" of $\mathbf{r}$ and $\mathbf{r}^*$ have a reciprocal relationship. In the context of diffraction phenomena, it is conventional to call the space of $\mathbf{r}$ as the direct space (or real space); it corresponds to the space of the scattering material. The space of $\mathbf{r}^*$ is usually referred to as the reciprocal space, and corresponds to the space of the observed diffraction pattern. While this distinction between the "direct" and "reciprocal" spaces is clearly obvious from a physical point of view, mathematically speaking, the terminology is interchangeable, and the two spaces are mutually reciprocal to each other. However, *directions* in the space $\mathbf{r}$ and the space $\mathbf{r}^*$ are directly comparable, e.g., a scalar product $\mathbf{r} \cdot \mathbf{r}^*$ is equal to $rr^* \cos \theta$ where θ is the angle between the directions of $\mathbf{r}$ and $\mathbf{r}^*$. We shall come across the reciprocal properties of the spaces of $\mathbf{r}$ and $\mathbf{r}^*$ prominently in connection with the diffraction by a lattice array (discussed briefly in a later section in this chapter).

Fourier transform of a delta function

We have seen earlier that the Fourier transform of the electron-density distribution in an atom, assumed to be spherically symmetrical, reduces to the integral (14) and becomes solely a function of $|\mathbf{r}^*| = 2(\sin \theta)/\lambda$. It would be of interest to consider a few other types of functions $\rho(\mathbf{r})$ and obtain their Fourier transforms, using (16).

Let us first consider a Gaussian function and, for convenience, deal with the one-dimensional case of $\rho(x)$ defined by

$$\rho(x) = \frac{1}{k} \exp \left(-\frac{\pi x^2}{k^2} \right). \tag{20}$$

Substitution in (16) leads readily to the transform of the form

$$\rho(x^*) = \exp \left(-\pi k^2 x^{*2} \right). \tag{21}$$

Thus the Fourier transform of a Gaussian function is also another Gaussian function in the inverse space. It may, however, be noted that the width $(1/k)$ of the latter is reciprocal to the width (k) in the original space.

This can be extended further. It may be readily seen that the infinite integral

$$\int_{-\infty}^{+\infty} \frac{1}{k} \exp \left(-\frac{\pi x^2}{k^2} \right) dx \tag{22}$$

is unity, whatever be the value of k. Consider the function

$$\delta(x) = \lim_{k \to 0} \frac{1}{k} \exp \left(\frac{\pi x^2}{k^2} \right). \tag{23}$$

This function $\delta(x)$ has therefore the following properties:

$$\delta(x) = 0 \quad \text{for} \quad x \neq 0$$
$$\delta(x) = \infty \quad \text{for} \quad x = 0$$

and   (24)

$$\int_{-\infty}^{+\infty} \delta(x)\, dx = 1.$$

This is the Dirac delta function and physically it corresponds to an infinitely sharp line of unit weight at the origin, i.e., for which the integral is unity. Some of the important properties of the delta function, which follow readily from the definition (22), are

$$\int_{-\infty}^{+\infty} f(x)\, \delta(x - x_0)\, dx = f(x_0),$$ (25)

$$\int_{-\infty}^{+\infty} \delta(x)\, \delta(x - x_0)\, dx = \delta(x_0).$$ (26)

In particular, putting $f(x) = \exp(2\pi ixx^*)$ in (25), we have

$$\int_{-\infty}^{+\infty} \exp(2\pi ix^*x)\, \delta(x - x_0)\, dx = \exp(2\pi ix^*x_0).$$ (27)

This relation shows that the Fourier transform of a delta function at x_0 is equal to $\exp(2\pi ix^*x_0)$. In particular, a delta function at the origin has for its Fourier transform the value unity for all values of x^* from $-\infty$ to $+\infty$. From the point of view of physical interpretation, we might say, using relations (20) and (21), that the delta function, considered as a Gaussian function with zero width, has for its Fourier transform a Gaussian function with infinite width. We may also reverse the arguments in view of the reciprocity relations—that is, the Fourier transform of a function which is a constant equal to unity at all points in direct space is a delta function at the origin in the reciprocal space, and the Fourier transform of an exponential function of x, namely $\exp(2\pi ixx_0^*)$ is $\delta(x^* - x_0^*)$.

Although we have considered, for simplicity, one-dimensional functions, the above results are true for higher dimensions also, the only change required being that, in the various formulas, the quantity xx^* has to be replaced by the scalar product $\mathbf{r} \cdot \mathbf{r}^*$.

Effect of translation

Suppose we have a function $\rho(\mathbf{r})$ and another function $\rho_1(\mathbf{r})$ which is obtained from $\rho(\mathbf{r})$ by means of a simple translation through a vector $\mathbf{t}$, i.e., $\rho_1(\mathbf{r}) = \rho(\mathbf{r} - \mathbf{t})$. If $F(\mathbf{r}^*)$ is the Fourier transform of $\rho(\mathbf{r})$ and $F_1(\mathbf{r}^*)$ that of $\rho_1(\mathbf{r})$, we get, using (16), that

$$
\begin{aligned}
F_1(\mathbf{r}^*) &= \int \rho(\mathbf{r} - \mathbf{t}) \exp\left(2\pi i \mathbf{r}^* \cdot \mathbf{r}\right) dv_r \\
&= \exp\left(2\pi i \mathbf{r}^* \cdot \mathbf{t}\right) \int \rho(\mathbf{r} - \mathbf{t}) \exp\left(2\pi i \mathbf{r}^* \cdot (\mathbf{r} - \mathbf{t})\right) dv_r \\
&= \exp\left(2\pi i \mathbf{r}^* \cdot \mathbf{t}\right) \int \rho(\mathbf{r}) \exp\left(2\pi i \mathbf{r}^* \cdot \mathbf{r}\right) dv_r \\
&= \exp\left(2\pi i \mathbf{r}^* \cdot \mathbf{t}\right) F(\mathbf{r}^*).
\end{aligned}
\tag{28}
$$

Thus translation by a vector $\mathbf{t}$ in the direct space is equivalent to multiplying the Fourier transform by $\exp\left(2\pi i \mathbf{r}^* \cdot \mathbf{t}\right)$ in the reciprocal space. This purely exponential factor may be referred to as the phase factor, since no change in magnitude is involved for the Fourier transform.

It follows that the Fourier transform of a set of δ functions $f_j \delta(\mathbf{r} - \mathbf{r}_j)$ is

$$
F(\mathbf{r}^*) = \sum_j f_j \exp\left(2\pi i \mathbf{r}^* \cdot \mathbf{r}_j\right).
\tag{28a}
$$

We shall call f_j as the weight of the δ function at $\mathbf{r}_j$.

Fourier transform of a molecule

As an illustration of the application of the phase factor, we may obtain (4) for the Fourier transform of a molecule containing N atoms as follows. Suppose $\mathbf{r}_j (j = 1$ to $N)$ are the vectors specifying the positions of the N atoms, with respect to an arbitrarily chosen origin. If $f_j(\mathbf{r}^*)$ is the Fourier transform of the atom j, referred to its center as origin, then the Fourier transform of the atom at $\mathbf{r}_j$ is $f_j(\mathbf{r}^*) \exp\left(2\pi i \mathbf{r}^* \cdot \mathbf{r}_j\right)$. Hence

$$
F(\mathbf{r}^*) = \sum_{j=1}^{N} f_j(\mathbf{r}^*) \exp\left(2\pi i \mathbf{r}^* \cdot \mathbf{r}_j\right).
\tag{29}
$$

Equation (29) differs from (4) in that $f_j(\mathbf{r}^*)$ is a function of the scattering vector $\mathbf{r}^*$, as the atom is not considered to be a point scatterer. If the atom is spherically symmetrical, $f_j(\mathbf{r}^*) = f_j(r^*)$, and f_j depends only on the magnitude of $\mathbf{r}^*$, but not on its direction. We shall make this assumption throughout this monograph, except where it is specifically stated to be otherwise.

However, it is important to remember that f_j is a varying function of r^* [or of $(\sin \theta)/\lambda$]. Consequently, although only the atomic centers have to be specified in (29), the variation of f_j with r^* takes care of the finite size and the nature of the electron-density distribution in each atom. This particular method of replacing the electron distribution in a molecule (or a crystal structure in general) by a finite set of atoms of variable scattering power is one that is used extensively in the later chapters of this monograph.

Diffraction by a crystal structure

We shall use the general equation (16) for a Fourier transform to work out the nature of the diffraction pattern given by a repetitive structure of scattering matter, as in a crystal. The atoms in a crystal are arranged according to the symmetry of a three-dimensional structure with unit-cell translation vectors **a**, **b**, **c**. Inside the unit cell, the different atoms (j) occur at positions $\mathbf{r}_j$ given by

$$\mathbf{r}_j = x_j \mathbf{a} + y_j \mathbf{b} + z_j \mathbf{c}, \tag{30a}$$

with

$$0 \leqslant x_j, y_j, z_j < 1, \tag{30b}$$

where x_j, y_j, z_j are the fractional coordinates of the atoms in the unit cell. As shown in Appendix C dealing with the convolution of sets of delta functions, the electron-density distribution of the whole crystal can be considered to be the convolution of the electron density distribution within a unit cell and a set of points (consisting of three-dimensional Dirac delta functions of unit weight) located at the lattice points defined by

$$\mathbf{r}(n_1, n_2, n_3) = n_1 \mathbf{a} + n_2 \mathbf{b} + n_3 \mathbf{c} \tag{31}$$

(see Appendix B for the definition of the term "convolution"). By (A15) of Appendix B, which gives the Fourier transform of the convolution of two functions as the product of the Fourier transforms of the individual functions, it follows that the Fourier transform of the whole crystal structure is the product of (*a*) the Fourier transform of the contents of the unit cell, and (*b*) the Fourier transform of the lattice. The first of these has already been considered in the last subsection and is given by (29), where the summation $j = 1$ to N is now over the atoms contained in the unit cell. We shall consider in some detail the second function.

Suppose the crystal is a parallelepipedal block and contains N_1 cells along the a direction, N_2 along b, and N_3 along c. Taking the origin at one corner of the parallelepiped and labelling this unit cell by $n_1 = 0, n_2 = 0, n_3 = 0$, it is readily seen that the Fourier transform of the lattice, $F_L(\mathbf{r}^*)$, is

$$F_L(\mathbf{r}^*) = \sum_{n_1=0}^{N_1-1} \sum_{n_2=0}^{N_2-1} \sum_{n_3=0}^{N_3-1} \exp\left[2\pi i \mathbf{r}^* \cdot (n_1 \mathbf{a} + n_2 \mathbf{b} + n_3 \mathbf{c})\right]. \tag{32}$$

The Fourier transform of the contents of the crystal, $F_C(\mathbf{r}^*)$, is

$$F_C(\mathbf{r}^*) = F(\mathbf{r}^*) F_L(\mathbf{r}^*), \tag{33}$$

where $F(\mathbf{r}^*)$ is given by (29) and $F_L(\mathbf{r}^*)$ by (32). The N atoms considered in (29) are now the contents of a unit cell of the crystal, referred to a suitably chosen origin. The scattered intensity in a direction $\mathbf{r}^*$ is thus proportional to $|F(\mathbf{r}^*)|^2|F_L(\mathbf{r}^*)|^2$. We must therefore examine in particular the function $|F_L(\mathbf{r}^*)|^2$.

From (32), $F_L(\mathbf{r}^*)$ is seen to be the product of three terms Q_1, Q_2, Q_3, of the type

$$Q_1 = \sum_{n_1=0}^{N_1-1} \exp\left(2\pi i n_1 \mathbf{r}^* \cdot \mathbf{a}\right), \text{ etc.,} \tag{34a}$$

so that $|F_L(\mathbf{r}^*)|^2$ takes the form of a product $P_1 P_2 P_3$, where

$$P_1 = \left| \sum_{n_1=0}^{N_1-1} \exp\left(2\pi i n_1 \mathbf{r}^* \cdot \mathbf{a}\right) \right|^2, \text{ etc.} \tag{34b}$$

These summations are readily worked out and we may arrive at the result

$$|F_L(\mathbf{r}^*)|^2 = \frac{\sin^2(N_1 \pi \mathbf{r}^* \cdot \mathbf{a})}{\sin^2(\pi \mathbf{r}^* \cdot \mathbf{a})} \cdot \frac{\sin^2(N_2 \pi \mathbf{r}^* \cdot \mathbf{b})}{\sin^2(\pi \mathbf{r}^* \cdot \mathbf{b})} \cdot \frac{\sin^2(N_3 \pi \mathbf{r}^* \cdot \mathbf{c})}{\sin^2(\pi \mathbf{r}^* \cdot \mathbf{c})}. \tag{35}$$

When N_1, N_2, N_3 are large, each of the terms in the right-hand side of (35) has appreciable values only when the equations

$$\mathbf{r}^* \cdot \mathbf{a} = h, \quad \mathbf{r}^* \cdot \mathbf{b} = k, \quad \mathbf{r}^* \cdot \mathbf{c} = l \tag{36}$$

(where h, k, l are integers) are approximately satisfied. The solution of the three equations in (36) can be shown to be given by[†]

$$\mathbf{r}^* = h\mathbf{a}^* + k\mathbf{b}^* + l\mathbf{c}^* = \mathbf{H} \tag{37a}$$

say, where

$$\mathbf{a}^* = \frac{\mathbf{b} \times \mathbf{c}}{\mathbf{a} \cdot (\mathbf{b} \times \mathbf{c})}, \quad \mathbf{b}^* = \frac{\mathbf{c} \times \mathbf{a}}{\mathbf{a} \cdot (\mathbf{b} \times \mathbf{c})}, \quad \mathbf{c}^* = \frac{\mathbf{a} \times \mathbf{b}}{\mathbf{a} \cdot (\mathbf{b} \times \mathbf{c})}. \tag{37b}$$

Also, it then follows that, when $N_1, N_2, N_3 \to \infty$, the function $|F_L(\mathbf{r}^*)|^2$ is given by

$$|F_L(\mathbf{r}^*)|^2 = N_1 N_2 N_3 \, \delta^{(3)}\,(\mathbf{r}^* - \mathbf{H}), \tag{38a}$$

where $\delta^{(3)}$ is the three-dimensional δ function. (See Appendix D for proof.) Therefore, we have, from (33),

$$|F_C(\mathbf{r}^*)|^2 = N_1 N_2 N_3 |F(\mathbf{r}^*)|^2 \, \delta^{(3)}(\mathbf{r}^* - \mathbf{H}). \tag{38b}$$

Thus it is seen that the distribution of scattered intensity in different directions has δ-function-type maxima corresponding to $\mathbf{r}^* = \mathbf{H}$. It can be shown

---

[†] The solution is most readily tested by directly substituting from (37a) and (37b) into each of the equations in (36), when they are found to be satisfied.

that the observed intensity corresponding to each maximum in a practical experiment is proportional to the integral of the δ function over a volume of reciprocal space including the particular value $\mathbf{r}^* = \mathbf{H}$. This quantity, which is known as the integrated intensity, is thus proportional to

$$N_1 N_2 N_3 \int |F(\mathbf{r}^*)|^2 \, \delta^{(3)}(\mathbf{r}^* - \mathbf{H}) \, dv_{\mathbf{r}^*} = N_1 N_2 N_3 \, |F(\mathbf{H})|^2 \qquad (39)$$

at $\mathbf{r}^* = \mathbf{H}$, where $\mathbf{H}$ takes all the values given by (37a). (See reference 8 for a physical description of integrated intensity, in which is also discussed the magnitude of the constant relating the observed intensity to $N_1 N_2 N_3 |F(\mathbf{H})|^2$ for different types of experimental arrangements.) Hence the scattered intensity of a volume of the crystal, of size large compared with the unit cell, is proportional to its volume and to the quantity $|F(\mathbf{H})|^2$, namely the square of the structure amplitude of the unit cell corresponding to the diffraction maximum $\mathbf{H}$.

Reciprocal lattice and Fourier series

Equation (39) shows that the Fourier transform of a crystal has appreciable values only for directions of scattering specified by the diffraction vector $\mathbf{r}^*$ being equal to $\mathbf{H}(= h\mathbf{a}^* + k\mathbf{b}^* + l\mathbf{c}^*)$. Since h, k, l are integers, the termini of the vectors $\mathbf{H}$ for all values of h, k, l form an infinite lattice with unit translations $\mathbf{a}^*, \mathbf{b}^*, \mathbf{c}^*$. From (37b), the quantities $\mathbf{a}^*, \mathbf{b}^*, \mathbf{c}^*$ will be in Å^{-1}, if $\mathbf{a}$, $\mathbf{b}, \mathbf{c}$ are measured in Å in the space of the crystal. Hence the term "reciprocal lattice" seems to be appropriate for the lattice defined by all the vectors $\mathbf{H}$. (Note: Just as we had seen earlier that the spaces of $\mathbf{r}$ and $\mathbf{r}^*$ are mutually reciprocal, so also the direct lattice defined by (31), and the reciprocal lattice defined by (37a), are also mutually reciprocal.) If the reciprocal of the reciprocal lattice were now defined, by the relations

$$\mathbf{a}^{**} = \frac{(\mathbf{b}^* \times \mathbf{c}^*)}{\mathbf{a}^* \cdot (\mathbf{b}^* \times \mathbf{c}^*)}, \text{ etc.}, \qquad (40)$$

it can be readily verified that $\mathbf{a}^{**} \equiv \mathbf{a}$ and so on.

Thus in the reciprocal space of a crystal lattice the Fourier transform is nonzero only at points[†] specified by the reciprocal lattice $\mathbf{H}$. At these points, however, the intensity of the diffracted wave is proportional to $|F(\mathbf{H})|^2$, where $F(\mathbf{r}^*)$ is defined by (29). Therefore, as far as regular crystal diffraction

---

[†] While the reciprocal vector $\mathbf{r}^*$ is general, we shall be using $\mathbf{H}$ to denote more specifically a reciprocal lattice point.

is concerned, we need consider only the function $F(\mathbf{H})$. In view of its continuous use throughout this monograph, we may state it explicitly in the form

$$F(\mathbf{H}) = \sum_{j=1}^{N} f_j \exp\left(2\pi i \mathbf{H} \cdot \mathbf{r}_j\right), \tag{41a}$$

or

$$F(hkl) = \sum_{j=1}^{N} f_j \exp\left[2\pi i(hx_j + ky_j + lz_j)\right]. \tag{41b}$$

The following result is also obvious

$$F(\mathbf{H}) = \int F(\mathbf{r}^*) \, \delta^{(3)}(\mathbf{r}^* - \mathbf{H}) \, dv_{r^*}. \tag{42}$$

The equation (41a) can also be written in terms of the continuous electron density $\rho(\mathbf{r})$ in the volume of the unit cell, and it then takes the form

$$F(\mathbf{H}) = \int_V \rho(\mathbf{r}) \exp\left(2\pi i \mathbf{H} \cdot \mathbf{r}\right) dv_r, \tag{43}$$

where the symbol V denotes integration over the volume V of a unit cell.

The inverse of (43) has the form of a Fourier series (rather than a Fourier-transform integral), as may be seen from the brief proof in Appendix E. It takes the form of a summation

$$\rho(\mathbf{r}) = \frac{1}{V} \sum_{\mathbf{H}} F(\mathbf{H}) \exp\left(-2\pi i \mathbf{H} \cdot \mathbf{r}\right), \tag{44a}$$

or

$$\rho(xyz) = \frac{1}{V} \sum_{h,k,l=-\infty}^{+\infty} F(hkl) \exp\left[-2\pi i(hx + ky + lz)\right]. \tag{44b}$$

It can be readily verified that (44) gives, not only the electron-density distribution in the unit cell at the origin, but also an identical distribution in every unit cell of the crystal. This is the Fourier series representation of electron density in a crystal.

Equations (41) and (43) are the fundamental equations that we will use throughout this monograph. Of these, the latter is quite general and is in terms of the continuous electron density in the unit cell. The former, however, uses the idea of the electron density being allocated to particular atoms. For convenience of reference we may write the exact inverse of (44b) in the explicit form

$$F(hkl) = \int_V \rho(xyz) \exp\left[2\pi i(hx + ky + lz)\right] dv_r, \tag{45}$$

where V denotes the volume of the unit cell.

Correspondingly, the left-hand side of (44) is also expressible in terms of the atoms j ($=1$ to N) and their scattering factors f_j (as a function of **H**). We shall consider this aspect in a later section.

Relation to experiment

It can be readily shown that the vector **H** of the reciprocal lattice is perpendicular to the set of lattice planes having the indices (hkl) and that its magnitude is equal to the reciprocal of the spacing of these planes. This is not proved here as it is beyond the scope of this monograph. It also follows from (3b), (7), and Fig. 1b that the diffraction maximum corresponds to a specular reflection from the lattice planes (hkl) of spacing $d(hkl)$ and that

$$\frac{1}{d(hkl)} = \frac{2 \sin \theta(hkl)}{\lambda}$$

or

$$\lambda = 2d(hkl) \sin [\theta(hkl)], \tag{46}$$

the well-known Bragg law.

As already mentioned, the intensity of an x-ray reflection hkl is proportional to $|F(hkl)|^2$. The exact scale factor connecting $|F(hkl)|^2$ and the measured intensity in an experiment can be worked out,[2,20] but we shall not be dealing with this here. We assume that the values of $|F(hkl)|^2$ are available and we shall consider the methods by which the crystal structure, or the electron-density distribution in the unit cell, may be obtained from these data, by means of Fourier methods. This leads us to a fundamental problem in crystallography, namely the so-called phase problem.

The phase problem

The relation (44) tells us that full information regarding the structure factors $F(\mathbf{H})$, or $F(hkl)$, will give us the structure $\rho(\mathbf{r})$ of the crystal. The structure factor is a complex quantity and therefore it can be written in the form

$$F(\mathbf{H}) = A(\mathbf{H}) + iB(\mathbf{H}). \tag{47a}$$

The real and imaginary parts, A and B, of F are shown in an Argand diagram (Fig. 3). Representing the phase of $F(\mathbf{H})$ by $\phi(\mathbf{H})$, also shown in Fig. 3, we have

$$F(\mathbf{H}) = |F(\mathbf{H})| \exp [i\phi(\mathbf{H})], \tag{47b}$$

where

$$|F(\mathbf{H})|^2 = A^2(\mathbf{H}) + B^2(\mathbf{H}), \tag{48}$$

and

$$\tan[\phi(\mathbf{H})] = \frac{B(\mathbf{H})}{A(\mathbf{H})}. \tag{49}$$

The measured intensities only give the values of $|F(\mathbf{H})|^2$ and therefore, only $|F(\mathbf{H})|$ on the right-hand side of (48) can be obtained by experiment, but not the phase $\phi(\mathbf{H})$ of $F(\mathbf{H})$. Thus, from experiment, we obtain directly only the moduli, but not the phases, of the structure factors which have to be

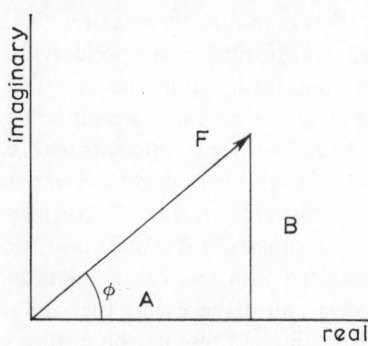

Fig. 3. Representation of the structure factor on the Argand diagram.

fed into a Fourier synthesis for obtaining the electron-density distribution. This difficulty is known as the phase problem and most of the theoretical approaches in crystallography are aimed toward obtaining these phases, either straightaway by the so-called "direct methods," or by making use of partial or other auxiliary information. We shall discuss in this monograph particularly the latter type of methods .For a general discussion of the phase problem, reference may be made to the standard texts in x-ray crystallography, e.g., references 2 and 9.

If the structure of the crystal contains centers of symmetry, i.e., to every atom with scattering factor f_j at $\mathbf{r}_j$, there is another with f_j at $-\mathbf{r}_j$, then the phase problem takes a simple form. Under these conditions, the imaginary components of $F(\mathbf{H})$ contributed by each such pair of atoms cancel each other, so that $B(\mathbf{H}) \equiv 0$. Therefore (48) takes the form

$$|F(\mathbf{H})|^2 = A^2, \tag{50}$$

giving

$$F(\mathbf{H}) = +A \quad \text{or} \quad -A \tag{51a}$$

or

$$F(\mathbf{H}) = A \quad \text{or} \quad A \exp(i\pi). \tag{51b}$$

Thus the general phase problem reduces to an ambiguity in the sign $(\pm)$ of $F(\mathbf{H})$ or to the question of which of the two possible values, namely 0 or π, the phase has. Throughout our discussions, we shall consider also separately this question of a centrosymmetric crystal, for which the phase problem and its solution takes much simpler forms.

Structure from Fourier synthesis

We have seen that a Fourier synthesis with coefficients $F(\mathbf{H}) = |F(\mathbf{H})| \exp [i\phi(\mathbf{H})]$ leads to the structure in direct space of which $F(\mathbf{H})$ are the structure factors, or the Fourier transforms, corresponding to the lattice points $\mathbf{H}$. Thus, if the structure contained within the unit cell consists of atoms of scattering factors f_j at positions $\mathbf{r}_j$, these would be recovered in the Fourier synthesis—ideally if all the Fourier coefficients with h, k, l up to infinity are included. We therefore call this structure, corresponding to the synthesis obtained with $F(\mathbf{H})$ as coefficients, as the "F structure" and the corresponding synthesis as the "F synthesis." We shall also use the term "strength" for the quantity f_j associated with atom j. As mentioned earlier, f_j is a function of $\mathbf{H}$, or of $(\sin \theta)/\lambda$, and it represents the size and shape of the atom. The strength of an atom thus represents the nature of the electron-density distribution in the atom. Symbolically, we may write

$$f_j, \mathbf{r}_j (j = 1 \text{ to } N) \to F(\mathbf{H}), \tag{52a}$$

or

$$f_{Nj}, \mathbf{r}_{Nj} \to F_N(\mathbf{H}) \tag{52b}$$

in which the symbol N, representing the number of atoms, is included optionally for $f_j, \mathbf{r}_j$, and $F(\mathbf{H})$, in which case the condition $j = 1$ to N need not be explicitly stated. Vice versa,

$$F_N(\mathbf{H}) \to f_{Nj}, \mathbf{r}_{Nj}. \tag{53}$$

Thus a synthesis with $F_N(\mathbf{H})$ as coefficients will lead to a structure with atoms of strength f_{Nj} at $\mathbf{r}_{Nj}$. In view of the equivalence of the right- and left-hand side of (52) and (53) in the direct and reciprocal spaces, we may call the structure $f_{Nj}, \mathbf{r}_{Nj}$ as the "F_N structure."

We may generalize these definitions and consider the structure in direct space obtained from a Fourier synthesis

$$\sum_{\mathbf{H}} \Phi(\mathbf{H}) \exp(-2\pi i \mathbf{H} \cdot \mathbf{r}), \tag{54}$$

where $\Phi(\mathbf{H})$ is a function of $\mathbf{H}$. We shall call this synthesis as the " Φ synthesis" and the corresponding structure in direct space obtained from the synthesis as the " Φ structure." We shall be dealing extensively with syntheses in which Φ is some function of F_N (and of other quantities, notably F_P, where P denotes a partial set of atoms, $k = 1$ to P, within the full set, $j = 1$ to N). In order to illustrate the nature of such syntheses, we shall take a very simple example, namely the $\tilde{F}(\mathbf{H})$ synthesis, where $\tilde{F}(\mathbf{H})$ is the complex conjugate of $F(\mathbf{H})$.

$\tilde{F}$ synthesis and the inverse structure

Since $\tilde{F}(\mathbf{H}) = |F(\mathbf{H})| \exp [-i\phi(\mathbf{H})]$, while $F(\mathbf{H}) = |F(\mathbf{H})| \exp [+i\phi(\mathbf{H})]$, it is obvious from (41a) that, if $(f_j, \mathbf{r}_j) \to F(\mathbf{H})$, then $(f_j, -\mathbf{r}_j) \to \tilde{F}(\mathbf{H})$. Conversely also, if a Fourier synthesis $F_N(\mathbf{H})$ leads to a structure $f_{Nj}, +\mathbf{r}_{Nj}$, the synthesis $\tilde{F}_N(\mathbf{H})$ would lead to a structure $f_{Nj}, -\mathbf{r}_{Nj}$. We shall call the synthesis $\tilde{F}(\mathbf{H})$ the "conjugate synthesis" to $F(\mathbf{H})$ and the structure $(f_j, -\mathbf{r}_j)$ the "inverse structure," for it is obtained from the original structure by inversion at the origin.

Negative structure and the inverse negative structure

We shall now apply the same method to two other examples. Consider the structure $-f_{Nj}, \mathbf{r}_{Nj}$, which has atoms at the same positions as the original structure from which we started, but in which the atoms have negative strengths, whose magnitudes are exactly equal to the corresponding positive strengths in the original structure $F_N(\mathbf{H})$. We call this the "negative structure" of the original structure. It is obvious that the Fourier transform of the negative structure is $-F_N(\mathbf{H})$.

Combining the idea of the inverse structure and the negative structure, it is obvious that we may think of the structure $-\tilde{F}_N(\mathbf{H})$, which will have peaks of strength $-f_{Nj}$ at $-\mathbf{r}_{Nj}$. It would therefore be the "negative inverse structure" of the original structure.

Table 1   Inverse and negative structures and their Fourier transforms

	Structure		Fourier
Name	Strengths	Positions	transform
Original	f_{Nj}	$\mathbf{r}_{Nj}$	F_N
Inverse	f_{Nj}	$-\mathbf{r}_{Nj}$	$\tilde{F}_N$
Negative	$-f_{Nj}$	$\mathbf{r}_{Nj}$	$-F_N$
Negative inverse	$-f_{Nj}$	$-\mathbf{r}_{Nj}$	$-\tilde{F}_N$

Combining all these, we obtain the results shown in Table 1.

Sharpening functions

In the foregoing sections we have seen how a Fourier synthesis using $F(\mathbf{H})$ as coefficients leads to the structure composed of atoms at $\mathbf{r}_j$ with strengths f_j. The atomic scattering factor f_j that appears in the various equations falls off as $\mathbf{r}^*$ increases, as has been already discussed earlier. This is due to the fact that atoms have finite size—that is, the electron density $\rho(\mathbf{r})$ extends over a distance not negligible compared to the wavelength of the x-rays used for diffraction. In certain problems it is convenient to formulate the various relations for the case of point atoms, that is, atoms which are confined to a volume element small compared to the wavelength of the x-rays used. We could represent such point atoms as delta functions of weight Z_j at $\mathbf{r}_j$. The scattering factor of such point atoms will have no fall off with $\mathbf{r}^*$ in the reciprocal space. This is obvious, since the Fourier transform of a delta function has a constant value, in this case equal to Z_j, the total number of electrons, in the point atom j. It is possible to simulate such a point atom, at least as a first approximation, by modifying the observed intensities of an actual crystal.

Thus we may define a quantity

$$\hat{f}_j = \frac{f_j}{Z_j}, \tag{55}$$

where $\hat{f}_j$ is the scattering factor normalized to unity at $\mathbf{r}^* = 0$. It is also referred to as the shape factor, since on multiplying it by Z_j, the number of electrons in an atom, we obtain the actual scattering factor of the atom. For simplicity, if we assume that all the atoms in the structure are alike, it is seen that

$$\frac{F(\mathbf{H})}{\hat{f}_j} = \sum_j Z_j \exp\left(2\pi i \mathbf{H} \cdot \mathbf{r}_j\right) \tag{56}$$

Thus the coefficients $F(\mathbf{H})$ ordinarily used in a Fourier synthesis, when divided by $\hat{f}_j$ and used in a new synthesis, will lead to point atoms of strength Z_j at positions $\mathbf{r}_j$.

In actual practice, it is not possible to get an exact conversion to the point-atom case, and the reasons are as follows. First, relation (56) is valid only in the case of a structure in which all the atoms are alike. Since in any ordinary structure the atoms are not all likely to be alike, we can only obtain either a mean value of $\hat{f}$ for the different atoms, or the quantity $\left(\sum_j f_j / \sum_j Z_j\right)$, instead of (55). Although the shape factors $\hat{f}_j$ for individual atoms are closely similar, they are not exactly the same, and to this extent the use of such an effective factor in (45) will introduce an approximation. Second, the scattering factors

f_j referred to above are the ones corresponding to an atom at rest. In actual practice, the atoms in a crystal undergo thermal vibrations and consequently their Fourier transform is different from f_j. Theory shows that the effect of this is to alter the scattering factor f_j to

$$f_j \exp\left[-B_j\left(\frac{\sin^2\theta}{\lambda^2}\right)\right]$$

where B_j is the temperature factor. In fact, $B_j = 8\pi^2\overline{u_j^2}$, $\overline{u_j^2}$ being the mean square amplitude of vibration of the atom. The value of mean B_j can usually be estimated from the measured intensity data.[20] We may then take this into account and modify the left-hand side of (56) into

$$\frac{F(\mathbf{H})}{\hat{f}_j} \exp\left[+B_j\left(\frac{\sin^2\theta}{\lambda^2}\right)\right].$$

Inasmuch as the values of B_j are approximate (and can also vary from atom to atom), the corresponding synthesis could only yield an approximation to point atoms of weight Z_j at $\mathbf{r}_j$.

A more serious and important factor is the following. Consider the intensity of an x-ray reflection. We have, using (41a),

$$I(\mathbf{H}) = F(\mathbf{H})\tilde{F}(\mathbf{H}) \tag{57a}$$

$$= \sum_j f_j^2 + \sum_{j \neq k} f_j f_k \exp[2\pi i\mathbf{H} \cdot (\mathbf{r}_j - \mathbf{r}_k)]. \tag{57b}$$

If we consider a small region of $(\sin\theta)/\lambda$ in the reciprocal space within which f_j may be assumed to be constant, the mean value of the intensities reduces to

$$\langle I(\mathbf{H}) \rangle = \sum_j f_j^2, \tag{58}$$

since the second term on the right-hand side of (57b) reduces to zero on the average. This statement is strictly true when all the atoms in the structure are alike, and are distributed randomly in the unit cell. The conclusions which follow are, however, not materially altered even in the case of unequal atoms. Thus, for an actual crystal, the mean intensity of the x-ray reflections falls off as $\mathbf{H}$ increases. Consider now the modified function $I(\mathbf{H})/\hat{f}^2$, where $\hat{f}$ has the same meaning as mentioned above. This operation is obviously equivalent to modifying the structure factors by dividing them by $\hat{f}$; hence we readily obtain

$$\frac{I(\mathbf{H})}{\hat{f}^2} = \sum_{j=1}^N Z_j^2 + \sum_{j \neq k} Z_j Z_k \exp[2\pi i H \cdot (\mathbf{r}_j - \mathbf{r}_k)]. \tag{59}$$

Thus the mean value of these modified intensities will be equal to $\sum Z_j^2$, which is a constant. Theoretically, therefore, there will not be any fall off of these modified intensities with increase of **H**.

The effect of such a modification is now obvious. Since the observed data are only over a limited range, the modified data become truncated sharply in reciprocal space. The use of these coefficients in a Fourier series is thus effectively equivalent to using a terminated series in the Fourier analysis. This results in serious bad effects, such as the introduction of negative ripples.[1,9] In particular, some of the peaks in the Patterson function may disappear (see Ch. 2).

The general approaches to the problem of trying to improve the observed data so as to correspond to point–atom models are commonly referred to as sharpening methods. The use of $\hat{f}$ discussed above is only one example. In general the aim is to optimize the conditions under which a function properly chosen would yield the best approximation to the point-atom case. Quite generally, we may represent the modified sharpening function as $M(\mathbf{H})$, so that the sharpened intensities are

$$I_S(\mathbf{H}) = M(\mathbf{H}) \cdot I(\mathbf{H}). \tag{60}$$

The function considered above corresponds to taking $M(\mathbf{H}) = 1/[\hat{f}(\mathbf{H})]^2$ Suppose we take

$$M(\mathbf{H}) = \frac{1}{\langle I(\mathbf{H}) \rangle}. \tag{61}$$

The resultant quantity is the normalized intensity, denoted by

$$|E(\mathbf{H})|^2 = \frac{I(\mathbf{H})}{\langle I(\mathbf{H}) \rangle}. \tag{62}$$

Since $\langle |F(\mathbf{H})|^2 \rangle$ has a mean value $(\sum_j f_j^2)$, the mean value of $|E(\mathbf{H})|^2$ may be readily seen to be a constant, equal to unity. Thus the use of $|E(\mathbf{H})|^2$ is also equivalent to using point atoms, each atom j having a weight equal to $f_j/(\sum_j f_j^2)^{1/2}$. The normalized intensities $|E(\mathbf{H})|^2$ are commonly used in statistical analysis.

A number of other functions are available which are more specifically aimed at handling the Patterson function and its improvement. These can be written in the general form[1]

$$M(r^*) = (r^*)^Q \exp N(r^*)^\gamma, \tag{63}$$

where r^* is the magnitude of the reciprocal vector, Q is an integer and N and γ are positive variables.

One of the simplest of such functions is $M(\mathbf{H}) = (\sin^2 \theta)/\lambda^2$. This has the simple interpretation that it is the Fourier transform of the self-convolution of a function which is the gradient of the electron density.[7] Other such functions are based more or less on empirical grounds and we shall not discuss them here. For details, the original papers may be referred to.[1,15] Some of the more recent methods[16] involve the extension of the intensity data artificially by calculation, beyond the limit actually observed, by iterative procedures.

Throughout this monograph we shall restrict ourselves to the use of the actual structure factors and intensities, although the theory, in general, can be extended to the case of the modified functions corresponding to the point-atom case. Thus the Fourier synthesis of an actual structure will be the one corresponding to using $F(\mathbf{H})$ as Fourier coefficients, and as discussed earlier, we shall refer to it as the structure containing atoms of strength f_j at positions $\mathbf{r}_j$. The quantity denoted by the "strength" f_j represents not only the peak height at the atom, but also the electron-density distribution surrounding the peak at position $\mathbf{r}_j$. The symbol f_j in fact stands for $f_j(\mathbf{r}^*)$, or $f_j[(\sin\theta)/\lambda]$ and the atom at $\mathbf{r}_j$ has an electron-density distribution, which is the Fourier transform of $f_j(\mathbf{r}^*)$. The crystal structure in the unit cell is the superposition of the electron-density distributions surrounding the locations $\mathbf{r}_j$.

If all the atoms are alike (in their electron distributions) and have the same temperature factor, then the peak heights can be taken to be proportional to f_j (e.g., see Lipson and Cochran,[9] pages 150–159). The peak heights for various elements, as a function of the temperature factor, have been published recently in tabular form.[17] As we shall show later, the possible uncertainties in the values of peak heights, which are affected by the temperature factor, series termination, etc., can be overcome to a good extent by normalizing them—that is, by taking the ratios of the different peak heights to that of a particular atom or the mean of a chosen set of atoms in the structure.

We shall, however, use the term "weight" (Z_j) to denote a point atom. In this case, since the peak height of a point delta function is not finite, the integral of the electron density is taken, which is finite and equal to Z_j.

Appendix

A. General Fourier transforms

We shall give here some of the results concerning Fourier transforms. Detailed treatments of the subject are available in standard works such as those by Titchmarsh,[19] Sneddon,[18] and Campbell and Foster.[3] A discussion of the Fourier transform in special relation to diffraction problems and crystallography is given in International Tables,[10] Vol. II.

Consider a function $f(x)$ of a single variable x. Its Fourier transform $F(x^*)$ is defined by

$$F(x^*) = \int_{-\infty}^{+\infty} f(x) \exp\left(2\pi i x x^*\right) dx. \tag{A1}$$

It can be shown that the function $f(x)$ can be given in terms of $F(x^*)$ by the inverse relation

$$f(x) = \int_{-\infty}^{+\infty} F(x^*) \exp\left(-2\pi i x x^*\right) dx^*. \tag{A2}$$

The proof of this follows from Fourier's integral theorem, which states that, for a certain class of functions,

$$f(x) = \int_{-\infty}^{+\infty} \int_{-\infty}^{+\infty} f(x') \exp\left[2\pi i (x' - x)x^*\right] dx' \, dx^*. \tag{A3}$$

Thus, consider the integral on the right-hand side of (A2). Substituting for $F(x^*)$ from (A1), we find it to be equal to

$$\int_{-\infty}^{+\infty} \left[\int_{-\infty}^{+\infty} f(x') \exp\left(2\pi i x' x^*\right) dx' \right] \exp\left(-2\pi i x x^*\right) dx^*$$

$$= \int_{-\infty}^{+\infty} \int_{-\infty}^{+\infty} f(x') \exp\left[2\pi i (x' - x)x^*\right] dx' \, dx^*,$$

which, by relation (A3), reduced to $f(x)$.

We may write relations (A1) and (A2) symbolically as

$$F(x^*) = T[f(x)],$$

and $\tag{A4}$

$$f(x) = T^{-1}[F(x^*)],$$

where T is the operator symbol which stands for multiplication by the kernel $\exp\left(2\pi i x^* x\right)$ and integration over x or x^* as the case may be, and T^{-1} corresponds to the inverse transformation, i.e., the same operation with the kernel $\exp\left(-2\pi i x^* x\right)$. It is equally valid to define the operator T to be associated with the kernel $\exp\left(-2\pi i x^* x\right)$ in which case T^{-1} will correspond to using $\exp\left(2\pi i x^* x\right)$ as the kernel. What is important, however, is that $F(x^*)$ and $f(x)$ are related by a pair of inverse transformations and involve the use of opposite signs in the kernels. In fact, we can see readily that $T[f(x)]$ being $F(x^*)$, $T^2[f(x)] = T[F(x^*)] = f(-x)$ and not $f(x)$.

It is readily seen from equations (A1) and (A2) that, if $f(x)$ is real, $F(x^*) = \tilde{F}(-x^*)$, where $\tilde{F}$ stands for the complex conjugate of F. If we write

$$F(x^*) = A(x^*) + i \, B(x^*), \tag{A5}$$

$A(x^*)$ and $B(x^*)$ are even and odd functions, respectively, of x^* given by

$$A(x^*) = \int_{-\infty}^{+\infty} f(x) \cos{(2\pi x^* x)}\, dx, \tag{A6}$$

$$B(x^*) = \int_{-\infty}^{+\infty} f(x) \sin{(2\pi x^* x)}\, dx. \tag{A7}$$

If, in addition, $f(x)$ is even, it is clear from (A7) that $B(x^*) = 0$ and hence $F(x^*)$ becomes real. On the other hand, if $f(x)$ is an odd function, $A(x^*) = 0$ and $F(x^*)$ is an imaginary function.

Fourier cosine and sine transforms

Suppose the function $f(x)$ is defined only for $x > 0$. We may define another function $\phi(x)$ such that

$$\phi(x) = f(x) \quad \text{for} \quad x > 0$$

and

$$\phi(-x) = f(x) \quad \text{for} \quad x < 0.$$

ϕ is thus defined for the entire range $-\infty$ to $+\infty$ but is an even function. Its transform $\Phi(x^*)$ is thus given by

$$\begin{aligned}
\Phi(x^*) &= \int_{-\infty}^{+\infty} \phi(x) \cos{(2\pi x^* x)}\, dx \\
&= 2 \int_{0}^{\infty} f(x) \cos{(2\pi x^* x)}\, dx.
\end{aligned} \tag{A8}$$

The inverse relation is

$$\begin{aligned}
f(x) &= \int_{-\infty}^{+\infty} \Phi(x^*) \cos{(2\pi x^* x)}\, dx^* \\
&= 2 \int_{0}^{\infty} \Phi(x^*) \cos{(2\pi x^* x)}\, dx^*.
\end{aligned} \tag{A9}$$

If, on the other hand, we define a function $\psi(x)$ as an odd function, with $\psi(x) = f(x)$ for $x > 0$ and $\psi(-x) = -f(x)$ for $x < 0$, it is clear that the relations would be

$$\begin{aligned}
\Psi(x^*) &= \int_{-\infty}^{+\infty} \psi(x) \sin{(2\pi x^* x)}\, dx \\
&= 2 \int_{0}^{\infty} f(x) \sin{(2\pi x^* x)}\, dx,
\end{aligned} \tag{A10}$$

and

$$f(x) = 2 \int_0^\infty \Psi(x^*) \sin (2\pi x^* x) \, dx^* \tag{A11}$$

Thus any function, defined only for one of the ranges $x < 0$, or $x > 0$, can be represented by a cosine or a sine transform, and the inverse transformation is of the same form in either case. The choice of one or the other for practical purposes depends to a large extent on the behavior of the function $f(x)$ at $x = 0$. For instance, from considerations of continuity of the function defined from $-\infty$ to $+\infty$, the sine function is more convenient to handle if $f(x) = 0$ for $x = 0$, whereas if $f(x)$ is finite at $x = 0$ the cosine function is more convenient.

Fourier transforms of various types of functions are listed in standard works.[3,18,19] Of particular interest to us is the Fourier transform of a delta function which has already been discussed in the main text (22) to (27). So also it has been shown that the effect of translation, by say x_0, is to multiply the Fourier transform by $\exp (2\pi i x^* x_0)$. Thus the transform of $f(x - x_0)$ is $F(x^*) \exp (2\pi i x^* x_0)$ (see (28), Ch. 1). So also the transform of $F(x^* - x_0^*)$ is $f(x) \exp (-2\pi i x_0^* x)$.

B. Convolution integral and its Fourier transform

Consider the function $q(x)$ defined by the integral

$$q(x) = \int_{-\infty}^{+\infty} f(x')g(x - x') \, dx', \tag{A12}$$

This function is usually referred to as the convolution, or Faltung, of the two functions $f(x)$ and $g(x)$ (Faltung in German means "folding.") It may be written as $f(x) * g(x)$. It can be shown that the function $q(x)$ can equally well be represented by the integral

$$q(x) = \int_{-\infty}^{+\infty} f(x - x')g(x') \, dx', \tag{A13}$$

and the relation between $f(x)$ and $g(x)$ is symmetrical in forming their convolution. When we wish to emphasize the individual functions which are convolved, we may use the notation $f * g(x)$ for $q(x)$.

An important theorem concerning the Fourier transform of $q(x) = f * g(x)$ is that if $F(x^*)$ and $G(x^*)$ are the Fourier transforms of $f(x)$ and $g(x)$, respectively, then the Fourier transform $Q(x^*)$ of $q(x)$ is the product of the Fourier transforms of $f(x)$ and $g(x)$, i.e., it is equal to $F(x^*)G(x^*)$.

We may prove this as follows. By the definition of the Fourier transform,

Appendix

we have

$$Q(x^*) = \int_{-\infty}^{+\infty} q(x) \exp\left(2\pi i x^* x\right) dx$$

$$= \int_{-\infty}^{+\infty} \int_{-\infty}^{+\infty} f(x')g(x - x') \exp\left(2\pi i x^* x\right) dx \, dx'. \tag{A14}$$

Substituting $x - x' = x''$, we obtain, since $dx = dx''$,

$$Q(x^*) = \int_{-\infty}^{+\infty} \int_{-\infty}^{+\infty} f(x')g(x'') \exp\left[2\pi i x^*(x' + x'')\right] dx' \, dx''$$

$$= \int_{-\infty}^{+\infty} f(x') \exp\left(2\pi i x^* x'\right) dx' \int_{-\infty}^{+\infty} g(x'') \exp\left(2\pi i x^* x''\right) dx'' \tag{A15}$$

$$= F(x^*) \, G(x^*)$$

which is the result stated above. It will be seen that $Q(x^*)$ is just the product of $F(x^*)$ and $G(x^*)$, and this would follow from either of the definitions (A12) or (A13) of the convolution.

It may be noted that if we convolve $f(x)$ with $f(-x)$, the resulting function $p(x) = f(x) * f(-x)$ has the Fourier transform

$$P(x^*) = F(x^*) \, \tilde{F}(x^*) = |F(x^*)|^2. \tag{A16}$$

We consider this function in great detail Ch. 2.

C. Convolutions involving delta functions

From the definition (A13) of the convolution of two functions, it follows that the convolution of a function $f(x)$ with a δ function, $\delta(x - x_0)$, yields the function $f(x - x_0)$. Symbolically,

$$\delta(x - x_0) * f(x) = f(x - x_0), \tag{A17a}$$

i.e., the process is equivalent to shifting the origin to x_0.

Suppose now that we have a periodic function $\phi(x)$, of period a, such that $\phi(x + na) = \phi(x)$ for integral values of n going from $-\infty$ to $+\infty$. Then the function is completely specified by giving the value of $\phi(x)$ in the range $0 \leqslant x < a$, and stating the period a. The latter translational symmetry may also be specified by a linear lattice of δ functions, i.e., by the function

$$f_L(x) = \sum_{n=-\infty}^{+\infty} \delta(x - na). \tag{A17b}$$

If we denote the value of $\phi(x)$ in the range $x = 0$ to a by $f(x)$ (defined only for $x = 0$ to a), then it follows from (A17a) and (A17b) that

$$\phi(x) = f(x) * f_L(x), \tag{A18a}$$

or, the complete function $\phi(x)$ is the convolution of the piece $f(x)$ from 0 to a and the lattice of δ functions. Hence, if $F(x^*)$ is the Fourier transform of $f(x)$ and $F_L(x^*)$ that of $f_L(x)$, the Fourier transform of $\phi(x)$ is

$$\Phi(x^*) = F(x^*)\, F_L(x^*). \tag{A18b}$$

The three-dimensional analog of the above was considered earlier and leads to the Fourier transform relation given in (33).

We shall now consider the convolution of functions composed of sets of δ functions. Suppose that the function $f(x)$ consists of a sum of δ functions, of weights f_i at x_i, i.e.,

$$f(x) = \sum_{i=1}^{N_1} f_i\, \delta(x - x_i), \tag{A19a}$$

and, similarly, let

$$g(x) = \sum_{j=1}^{N_2} g_j\, \delta(x - x_j). \tag{A19b}$$

By making use of the definition (A12), or (A13), for $q(x) = f * g(x)$, it follows that $q(x)$ will have infinitely sharp peaks, of weights $f_i g_j$ at $(x_i + x_j)$, or

$$q(x) = \sum_{i=1}^{N_1} \sum_{j=1}^{N_2} f_i g_j\, \delta[x - (x_i + x_j)]. \tag{A19c}$$

This result is fundamental to all the discussions in this monograph. The convolution has $N_1 N_2$ peaks, in general all different, whose weights are the *products* of those of the individual peaks in the two original functions, and whose positions are located at the *sums* of the x coordinates of the corresponding peaks in the original two functions. We shall consider a corresponding result for atoms in a crystal structure in Ch. 2.

D. *Proof of the grating formula*

The right-hand side of (35) is a product of three terms. A typical one may be written in the form $\sin^2 (N_1 \pi h')/\sin^2 (\pi h')$, where $h' = \mathbf{r}^* \cdot \mathbf{a}$ is a continuous variable. It is readily seen that, when N_1 is large, this function has a large value, $= N_1^2$, for $h' = 0, 1, 2, \ldots$ (h, say, in general) and that it then rapidly falls to zero for $h' = h \pm 1/N_1$ and that, outside these ranges, it has a value $< 1/N_1$ times the peak value for $h' = h$. Hence, for sufficiently large N_1, the function $\sin^2 (N_1 \pi h')/\sin^2 (\pi h')$ may be replaced by a set of delta-function-like singularities of the form $f(h') = \sin^2 (N_1 \pi h')/(\pi h')^2$. In the limit when $N_1 \to \infty$, this function has the property of a delta function, for $f(h') \to \infty$ at $h' = h$, and the half-width of the peak at $h' = h$ is $1/N_1$, which tends to zero. However, the weight of the peak is not unity, but N_1, as may be seen from the value of the integral in (A20).

$$\int_{-\infty}^{+\infty} \frac{\sin^2{(N_1 \pi h')}}{(\pi h')^2} \, dh' = \frac{N_1}{\pi} \int_{-\infty}^{+\infty} \frac{\sin^2{(N_1 \pi h')}}{(N_1 \pi h')^2} \, d(N_1 \pi h')$$

$$= \frac{N_1}{\pi} \int_{-\infty}^{+\infty} \frac{\sin^2{y}}{y^2} \, dy = \frac{N_1}{\pi} \pi = N_1. \qquad \text{(A20)}$$

Thus, when $N_1 \to \infty$, $\sin^2{(N_1 \mathbf{r^*} \cdot \mathbf{a})}/\sin^2{(\pi \mathbf{r^*} \cdot \mathbf{a})} = N_1 \, \delta(\mathbf{r^*} \cdot \mathbf{a} - h)$. Similarly treating the other two terms in (35), we finally obtain

$$|F_L(\mathbf{r^*})|^2 = N_1 N_2 N_3 \, \delta(\mathbf{r^*} \cdot \mathbf{a} - h) \, \delta(\mathbf{r^*} \cdot \mathbf{b} - k) \, \delta(\mathbf{r^*} \cdot \mathbf{c} - l). \qquad \text{(A21)}$$

The product of the three delta functions is clearly the three-dimensional delta function $\delta^{(3)}(\mathbf{r^*} - \mathbf{H})$, where $\mathbf{H} = h\mathbf{a^*} + k\mathbf{b^*} + l\mathbf{c^*}$. This leads to (38).

E. Periodic functions, their Fourier transform and inversion in terms of Fourier series

The Fourier series representation of periodic functions is well known and is treated, for example, by Sneddon[18] and in International Tables,[10] Vol. II. Here we shall give a formulation of it which is directly applicable to crystallography. We shall derive this as a particular case of the Fourier transform theory as given in Appendix A.

Suppose a function $f(x)$ is periodic, of period a, such that

$$f(x + na) = f(x), \quad n = \text{an integer}, \quad -\infty < n < +\infty. \qquad \text{(A22)}$$

Then, it readily follows from (A1) that, for $x^* \neq h/a(h = \text{an integer})$, its Fourier transform, which we shall denote by $\Phi(x^*)$, is zero. However, when x^* is exactly equal to a multiple of $1/a$, say h/a, then

$$\Phi\!\left(\frac{h}{a}\right) = \sum_{-\infty}^{+\infty} \int_0^a f(x) \exp\left[\frac{2\pi i(hx + na)}{a}\right] dx$$

$$= \left[\int_0^a f(x) \exp\left(\frac{2\pi i hx}{a}\right) dx\right]\left[\sum_{-\infty}^{+\infty} \exp{(2\pi i nh)}\right]. \qquad \text{(A23)}$$

The quantity becomes infinite, because every term in the second sum is unity and it does not converge. However, these infinities at $x^* = h/a$ are of the δ-function type, as may be seen from (A24) below:

$$\Phi(x^*) = \left[\int_0^a f(x) \exp{(2\pi i x^* x)} \, dx\right]\left[\sum_{-\infty}^{+\infty} \exp{(2\pi i x^* na)}\right], \qquad \text{(A24)}$$

in which the second term in the right-hand side of (A24) can be shown to be equal to

$$k(x^*) = \mathop{Lt}_{N \to \infty} \frac{\sin(\pi N x^* a)}{\sin(\pi x^* a)} . \tag{A25}$$

This function $k(x^*)$ has infinitely sharp peaks at $x^* = h/a$, but the weight of each peak is $1/a$, as may be seen from the value of the integral

$$\mathop{Lt}_{N \to \infty} \int_{-\infty}^{+\infty} \frac{\sin(\pi N x^* a)}{\sin(\pi x^* a)} dx^* = \mathop{Lt}_{N \to \infty} \frac{1}{\pi a} \int_{-\infty}^{+\infty} \frac{\sin(\pi N x^* a)}{(\pi N x^* a)} d(\pi N x^* a) = \frac{1}{a},$$

$$\tag{A26}$$

so that

$$k(x^*) = \frac{1}{a} \delta \left(x^* - \frac{h}{a} \right). \tag{A27}$$

In view of the above properties, it is convenient to consider the first term of (A24), which is the Fourier transform of the repeating unit, as the representative Fourier transform of a periodic function. Denoting this by $F(x^*)$, we have

$$F(x^*) = \int_0^a f(x) \exp(2\pi i x^* x) \, dx, \tag{A28}$$

and it follows that

$$\Phi(x^*) = \frac{1}{a} \sum_{h=-\infty}^{+\infty} F(x^*) \delta \left(x^* - \frac{h}{a} \right). \tag{A29}$$

Now, from the general theory of Fourier transforms given in Appendix A, the function $f(x)$ is the inverse Fourier transform of $\Phi(x^*)$, or

$$f(x) = \int_{-\infty}^{+\infty} \Phi(x^*) \exp(-2\pi i x^* x) \, dx^* \tag{A30}$$

Substituting for $\Phi(x^*)$ from (A29), we have

$$f(x) = \frac{1}{a} \int_{-\infty}^{+\infty} \sum_{h=-\infty}^{+\infty} F(x^*) \exp(-2\pi i x^* x) \delta \left(x^* - \frac{h}{a} \right) dx^*$$

$$= \frac{1}{a} \sum_{h=-\infty}^{+\infty} \int_{-\infty}^{+\infty} F(x^*) \exp(-2\pi i x^* x) \delta \left(x^* - \frac{h}{a} \right) dx^* \tag{A31}$$

$$= \frac{1}{a} \sum_{h=-\infty}^{+\infty} F(h/a) \exp \left(\frac{-2\pi i h x}{a} \right).$$

Thus, $f(x)$ is obtained as a summation of an exponential series, whose coefficients are $F(h/a)$. It is usual, in crystallographic literature, to denote $F(h/a)$ by $F(h)$, whose value, from (A28) is

$$F(h) = \int_0^a f(x) \exp\left(\frac{2\pi ihx}{a}\right) dx. \tag{A32}$$

The inverse relation is

$$f(x) = \frac{1}{a} \sum_{h=-\infty}^{+\infty} F(h) \exp\left(\frac{-2\pi ihx}{a}\right) \tag{A33}$$

Equations (43) and (44) of the main text are the three-dimensional analogs of (A32) and (A33), using *fractional* coordinates x, y, z.

F. Spherical distributions

We shall give a brief derivation of (14) and some of the consequences arising from it. We have, from (13), that

$$f(\mathbf{r}^*) = \int \rho(\mathbf{r}) \exp\left(2\pi i \mathbf{r}^* \cdot \mathbf{r}\right) dv_r.$$

Suppose $\rho(\mathbf{r})$ is spherically symmetrical, i.e., that it is a function of r only. Then we may represent $\mathbf{r}$ and $\mathbf{r}^*$ in spherical polar coordinates (r, θ, ϕ) and (r^*, Θ, Φ), respectively. We then obtain

$$f(r^*, \Theta, \Phi) = \int_0^\infty \int_0^{\pi/2} \int_0^{\pi/2} \rho(r, \theta, \phi) \exp\left\{2\pi i r^* r[\cos\theta \cos\Theta\right.$$
$$\left. + \sin\theta \sin\Theta \cos(\phi - \Phi)]\right\} r^2 \sin\theta \, dr \, d\theta \, d\phi. \tag{A34}$$

If $\rho(r, \theta, \phi) = \rho(r)$, this can be integrated, and $F(r^*, \Theta, \Phi)$ is found to be a function of r^* only, namely

$$f(r^*) = \int_0^\infty 4\pi r^2 \rho(r) \frac{\sin(2\pi r r^*)}{(2\pi r r^*)} dr, \tag{A35}$$

which is identical with (14). Thus $f(r^*)$ is also spherically symmetrical. Vice versa, we also have the equation

$$\rho(r) = \int_0^\infty 4\pi r^2 f(r^*) \frac{\sin(2\pi r r^*)}{(2\pi r r^*)} dr^*. \tag{A36}$$

An interesting consequence of the above two equations is that, when $r^* = 0$,

$$f(0) = \int_0^\infty 4\pi r^2 \rho(r) \, dr = Z, \tag{A37}$$

where Z, in the case of atoms scattering x-rays, is just the total number of electrons. Vice versa, we also have

$$\rho(0) = \int_0^\infty 4\pi r^2 f(r^*) \, dr^* \qquad (A38)$$

or the peak electron density (at the origin) is equal to the integral of $F(r^*)$ over reciprocal space.

References

[1] S. Abrahamsson and E. N. Maslen. *The use of diverging modification in the solution of three-dimensional Patterson synthesis*, Z. Kristallogr. **118** (1963) 1–32.

[2] M. J. Buerger. *Crystal-structure analysis*. (Wiley, New York, 1960).

[3] G. A. Campbell and R. M. Foster. *Fourier integrals for practical calculations*. Bell telephone system tech. publ., Monograph B584 (1931).

[4] B. Dawson. *Aspherical atomic scattering factors in crystal structure refinement I. Coordinate and thermal motion effects in a model centrosymmetric system*. Acta Crystallogr. **17** (1964) 990–996.

[5] B. Dawson. *Aspherical atomic scattering factors for some light atoms in sp^3, sp^2 and sp hybrid valence state approximations*, Acta Crystallogr. **17** (1964) 997–1009.

[6] A. J. Freeman. *Atomic scattering factors for spherical and aspherical charge distributions*. Acta Crystallogr. **12** (1959) 261–271.

[7] Robert A. Jacobson, Jeffrey A. Wunderlich, and William N. Lipscomb. *The crystal and molecular structure of cellobiose*. Acta Crystallogr. **14** (1961) 598–607.

[8] R. W. James. *The optical principles of the diffraction of X-rays*. (G. Bell and Sons, London, 1948).

[9] H. Lipson and W. Cochran. *The crystalline state, Vol. III. The determination of crystal structures*. (G. Bell and Sons, London, 1953) 291–298.

[10] Caroline H. MacGillavry and Gerard D. Rieck (Ed). *International tables for x-ray crystallography, Vol. III*, (Kynoch Press, Birmingham, England, 1962).

[11] R. McWeeny. *X-ray scattering by aggregates of bonded atoms. I. Analytical approximations in single-atom scattering*. Acta Crystallogr. **4** (1951) 513–519.

[12] R. McWeeny. *X-ray scattering by aggregates of bonded atoms. II. The effects of the bonds with an application to H_2*. Acta Crystallogr. **5** (1952) 463–468.

[13] R. McWeeny. *X-ray scattering by aggregates of bonded atoms. III. The bond scattering factor. Simple methods of approximation in the general case*. Acta Crystallogr. **6** (1953) 631–637.

[14] R. McWeeny. *X-ray scattering by aggregates of bonded atoms. IV. Application to carbon atom*. Acta Crystallogr. **7** (1954) 180–186.

[15] R. Ramachandra Ayyar. *Tests of Patterson sharpening functions*. Ind. J. Pure Appl. Phys. **5** (1967) 382–386.

[16] S. Raman and J. Lawrence Katz. *An analytical method of obtaining sharpened Patterson functions*. Z. Kristallogr. **124** (1967) 43–63.

Appendix

[17] Tosio Sakurai. *Peak heights of electron density and Patterson function in the crystal.* Acta Crystallogr. **23** (1967) 862–865.
[18] I. N. Sneddon. *Fourier transforms.* (McGraw Hill, New York, 1951).
[19] E. C. Titchmarsh. *Theory of Fourier integrals.* (Cambridge University Press, 1937).
[20] A. J. C. Wilson. *Determination of absolute from relative x-ray intensity data.* Nature **150** (1942) 152.

2

Patterson and Patterson-type syntheses

Convolution of two structures

The convolution $q(\mathbf{r})$ of two structures $f(\mathbf{r})$ and $g(\mathbf{r})$ in three dimensions may be written in either of the two forms below, following the equations given in the Appendix B to Ch. 1.

$$q(\mathbf{r}) = \int f * g(\mathbf{r})$$

$$= \int f(\mathbf{r} - \mathbf{r}')g(\mathbf{r}') \, dv_{r'} \tag{1a}$$

$$= \int f(\mathbf{r}')g(\mathbf{r} - \mathbf{r}') \, dv_{r'}. \tag{1b}$$

The Fourier transform $Q(\mathbf{r}^*)$ of $q(\mathbf{r})$ is then obviously given by

$$Q(\mathbf{r}^*) = F(\mathbf{r}^*) \, G(\mathbf{r}^*). \tag{2}$$

When $f(\mathbf{r})$ and $g(\mathbf{r})$ are made up of sums of three-dimensional delta functions, we have, analogous to (A16)–(A18) of Ch. 1, the following formulas. If

$$f(\mathbf{r}) = \sum_{i=1}^{N_1} f_i \, \delta^{(3)}(\mathbf{r} - \mathbf{r}_i), \tag{3a}$$

$$g(\mathbf{r}) = \sum_{j=1}^{N_2} g_j \, \delta^{(3)}(\mathbf{r} - \mathbf{r}_j), \tag{3b}$$

and if $q(\mathbf{r}) = f * g(\mathbf{r})$, then

$$q(\mathbf{r}) = \sum_{i=1}^{N_1} \sum_{j=1}^{N_2} f_i g_j \, \delta^{(3)}(\mathbf{r} - \mathbf{r}_i - \mathbf{r}_j). \tag{4}$$

34

We shall derive the result (4) from (3a) and (3b) by means of the Fourier transform relation (2) to illustrate a general method of approach, which will be adopted largely in this monograph. The Fourier transform of $f(\mathbf{r})$ and $g(\mathbf{r})$ are clearly

$$F(\mathbf{r}^*) = \sum_{i=1}^{N_1} f_i \exp(2\pi i \mathbf{r}^* \cdot \mathbf{r}_i), \qquad (5a)$$

and

$$G(\mathbf{r}^*) = \sum_{i=1}^{N_2} g_j \exp(2\pi i \mathbf{r}^* \cdot \mathbf{r}_j). \qquad (5b)$$

We have, therefore, from (2)

$$Q(\mathbf{r}^*) = \sum_{i=1}^{N_1} \sum_{j=1}^{N_2} f_i g_j \exp[2\pi i \mathbf{r}^* \cdot (\mathbf{r}_i + \mathbf{r}_j)] \qquad (6)$$

and the inverse Fourier transform of this gives (4).

Thus, we see that the convolution of two δ-function structures $\mathscr{S}_1$ and $\mathscr{S}_2$ defined by (7a) and (7b) below, is the structure $\mathscr{C}_{12}$ given by (8) below.

$$\mathscr{S}_1 : f_{N_1 i} \quad \text{at} \quad \mathbf{r}_{N_1 i}, \quad i = 1 \quad \text{to} \quad N_1, \qquad (7a)$$

$$\mathscr{S}_2 : f_{N_2 j} \quad \text{at} \quad \mathbf{r}_{N_2 j}, \quad j = 1 \quad \text{to} \quad N_2, \qquad (7b)^\dagger$$

$$\mathscr{C}_{12} = \mathscr{S}_1 * \mathscr{S}_2 : f_{N_1 i} f_{N_2 j} \quad \text{at} \quad \mathbf{r}_{N_1 i} + \mathbf{r}_{N_2 j}. \qquad (8)^\dagger$$

This result is quite general for any such structures in direct space. However, we will be mainly interested in repetitive structures in a crystal, and further, we are interested in electron-density distributions which are the sum of "atoms" of finite size rather than delta-function singularities. Both these aspects can be taken into consideration as shown below.

As mentioned in Ch. 1, the restriction to structures within the unit cell of a crystal and its repetitive nature leads to the result that we need consider only the values of $F(\mathbf{r}^*)$ at $\mathbf{r}^* = \mathbf{H}$, the points of the appropriate reciprocal lattice. The occurrence of atoms can be simulated by taking the scattering factor of an atom i at $\mathbf{r}_i$ to be $f_i(\mathbf{H})$, the function f_i having the appropriate variation with H or $(\sin\theta)/\lambda$. With this redefinition of the f's occurring in (7) and (8), the two equations continue to hold for a crystal structure also. The only difference is that $F(\mathbf{r}^*)$, $G(\mathbf{r}^*)$, and $Q(\mathbf{r}^*)$ in (5) and (6) are zero, except for $\mathbf{r}^* = \mathbf{H}$, a reciprocal-lattice vector. Further, (7a) and (7b) would mean that the electron density at any point in the unit cell is the sum of the electron densities at the point of the appropriate distributions $\rho_{Ni}(\mathbf{r} - \mathbf{r}_{Ni})$

---

† In order to agree with the later usage, we use $f_{N_2 j}$ rather than g_j for the weights of the delta functions in $\mathscr{S}_2$ in these equations.

of the various atoms. Apart from this way of looking at the electron-density distribution, there are no approximations involved in the process.

The structure $\mathscr{C}_{12}$, in (8), consists of atoms of strengths $f_i f_j$ at $\mathbf{r}_i + \mathbf{r}_j$. To fix our attention, let us suppose that the atoms i and j are of the same type and that $f_i = f_j = f(\mathbf{r}^*)$. Then $f_i f_j = f^2(\mathbf{r}^*)$; this declines much more rapidly than $f(\mathbf{r}^*)$. Consequently the peak in direct space at $\mathbf{r}_i + \mathbf{r}_j$ in $\mathscr{C}_{12}$ is less sharp than the corresponding peaks at $\mathbf{r}_i$ and $\mathbf{r}_j$ in $\mathscr{S}_1$ and $\mathscr{S}_2$, respectively. On the other hand, the peak value, which is given by $\int_0^\infty f^2(r^*)\, 4\pi r^{*2}\, dr^*$, ((A38) of Ch. 1), is larger in $\mathscr{C}_{12}$ than in $\mathscr{S}_1$ or $\mathscr{S}_2$. The extension to the case when $f_i \neq f_j$ is obvious. Using the definition of the structure F_N given in Ch. 1, we have the following results, which are of great interest in relation to the succeeding chapters.

(1) If $\mathscr{S}_1 \equiv$ structure F_{N_1} and $\mathscr{S}_2 \equiv$ structure F_{N_2}, then the convolution $\mathscr{C}_{12}$ of $\mathscr{S}_1$ and $\mathscr{S}_2$ is the structure $F_{N_1} F_{N_2}$, and (2) the peaks of the structure $F_{N_1} F_{N_2}$ have strengths $f_{N_1 i} f_{N_2 j}$ occurring at positions $\mathbf{r}_{N_1 i} + \mathbf{r}_{N_2 j}$. For the above results to be valid for a crystal structure, the two functions F_{N_1} and F_{N_2} must be defined over the same set of reciprocal-lattice points. In others words, the structures $\mathscr{S}_1$ and $\mathscr{S}_2$ must have the same unit cell. Thus, they may be two different sets of atoms in the unit cell, or they may be the same set of atoms, or one set may be related to the other by some formula, or one may be a part of the other. A variety of results follow under these various conditions. We shall discuss some of these in this chapter.

Patterson structure

It was first shown by Patterson[2,3] that a Fourier synthesis performed with the intensities $I(\mathbf{H})$ of the x-ray reflections as coefficients contain high values corresponding to the interatomic vectors between the atoms in the structure. This could be seen readily as follows. Since $I(\mathbf{H}) = F(\mathbf{H})\tilde{F}(\mathbf{H})$, the Patterson is the convolution of the structure $F(\mathbf{H})$, consisting of the atoms in the unit cell, with its inverse structure $\tilde{F}(\mathbf{H})$. We have already seen in Ch. 1 that if the structure $F \equiv (f_{Nj}, \mathbf{r}_{Nj})$, then the inverse structure $\tilde{F}(\mathbf{H})$ contains peaks of strength f_{Nj} at positions $-\mathbf{r}_{Nj}$, which are inverse to the positions of the peaks in the structure $F(\mathbf{H})$. Thus the Patterson synthesis should contain peaks of strengths $f_{Ni} f_{Nj}$ at positions $(\mathbf{r}_{Ni} - \mathbf{r}_{Nj})$, which are the interatomic vectors.

In view of its importance in structure analysis and in our later discussions, it will be useful to analyze the properties of the Patterson function a little more in detail.

Noncentrosymmetric structure[4,5]

Let us consider a structure containing N atoms defined by position vectors $\mathbf{r}_j (j = 1$ to $N)$.[†] The Patterson would contain peaks of strengths $f_i f_j$ at

---

[†] In this and the next subsection, we shall omit the subscript N, as it is the same throughout.

$(\mathbf{r}_i - \mathbf{r}_j)(i, j = 1$ to $N)$. It is obvious, since any vector $(\mathbf{r}_i - \mathbf{r}_j) = -(\mathbf{r}_j - \mathbf{r}_i)$, that the Patterson structure is centrosymmetric in its properties. Further, all the self vectors (i.e., those with $i = j$) totaling N in number give rise to a peak of strength

$$\sum_{j=1}^{N} f_j^2 = S_N^2 \qquad (9)$$

at the origin. The peaks for which $i \neq j$ are therefore $N(N - 1)$ in number, half of which, namely $N(N - 1)/2$, are related to the other half by a center of symmetry at the origin. These are summarized in Table 1.

Since the Patterson diagram is of fundamental importance both in the theory and applications of structure analysis, we shall introduce a notation for representing the vectors in this diagram, a method which is explored further in Ch. 3 and utilized in the later chapters. We indicate a vector from an atom i to an atom j in the original structure by a symbol $[ij]$. This symbol represents a vector $\mathbf{r}_j - \mathbf{r}_i$. We also use the curly brackets $\{ij\}$ to represent the set of all vectors (N^2 in number) with i, j going from 1 to N. The symbol $\{1i\}$ or $\{i1\}$ would represent N vectors of the types $\mathbf{r}_i - \mathbf{r}_1$ and $\mathbf{r}_1 - \mathbf{r}_i$, respectively. The vectors $\{ii\}$ are all equal to the null vector and terminate at the origin. Thus the Patterson consists of a peak of strength S_N^2 at the origin consisting of these N superposed peaks, and $N(N - 1)/2$ pairs of peaks $\{ij\}$ and $\{ji\}$ $(j > i)$ of strengths $f_i f_j$.

Centrosymmetric structures

When a structure possesses a center of symmetry, this obviously produces additional relations between the interatomic vectors and consequently the features of the Patterson are affected. Suppose that there are M atoms in the asymmetric part of the structure, each of strength f_j at positions $\mathbf{r}_j (j = 1$ to M, $M = N/2$). These are related by the center of inversion at the origin to another set of M atoms of strength $f_{j'}$ at $-\mathbf{r}_{j'}$ (denoted by $j' = 1$ to M). It is readily seen that, apart from the origin peak, which is of strength $2S_M^2 = S_N^2$, there will be two types of interactions in the centrosymmetric Patterson. The first type are those which may be termed "single peaks," which arise from vectors between symmetry-related like atoms and may be represented by the symbol $[jj']$ $(j, j' = 1$ to M) and $[j'j]$. There are $2M(=N)$ such peaks and they are of strength f_j^2. The second type of peaks are those between atoms not related by symmetry. In this category two identical vectors exist, namely of the type $[ij]$ and $[j'i']$ at $(\mathbf{r}_j - \mathbf{r}_i) \equiv (\mathbf{r}_{i'} - \mathbf{r}_{j'})$. The corresponding peaks are therefore double peaks in the Patterson and their strengths will be $2f_i f_j$. Table Ib gives the types of Patterson peaks for a centrosymmetric structure. For convenience, the double peaks themselves have been grouped under two categories, one of type $\mathbf{r}_j - \mathbf{r}_i$ and the other of type $\mathbf{r}_j + \mathbf{r}_i$, i.e., corresponding to vectors $[ij] \equiv [j'i']$ and $[ij'] \equiv [ji']$, respectively.

Patterson and Patterson-type syntheses

Table 1   Types of peaks in the Patterson diagram and in the
squared structure

(a) Noncentrosymmetric structure

Position	Patterson Strength	Number	Position	Squared structure Strength	Number
Origin	S_N^2	1	$2\mathbf{r}_j$	f_j^2	N
$\mathbf{r}_j - \mathbf{r}_k$	$f_j f_k$	$N(N-1)$	$\mathbf{r}_j + \mathbf{r}_k$	$2f_j f_k$	$N(N-1)/2$

(b) Centrosymmetric structure

Patterson and squared structure

Position	Strength	Number
Origin	$S_N^2 (N = 2M)$	1
$\pm 2\mathbf{r}_{Mj}$	f_{Mj}^2	N
$\pm(\mathbf{r}_{Mj} - \mathbf{r}_{Mk})$	$2f_{Mj}f_{Mk}$	$N\left(\dfrac{N}{2} - 1\right)$
$\pm(\mathbf{r}_{Mj} + \mathbf{r}_{Mk})$	$2f_{Mj}f_{Mk}$	$N\left(\dfrac{N}{2} - 1\right)$

Thus it is seen that the introduction of a center of symmetry decreases the total number of distinct peaks in the Patterson and this, in a sense, helps in the interpretation of the Patterson function. In fact, if it is possible to recognize a single peak, then it would be possible to apply any of the image-seeking procedures, to be discussed in Ch. 3, which would lead, in principle, to a complete solution of the structure of a centrosymmetric crystal from its Patterson in a simple way.

Squared structure

The Patterson uses $|F|^2 = F\tilde{F}$ as coefficients for a Fourier synthesis. Consider a synthesis using $F^2 = FF$ as coefficients. We shall call this the squared structure, a name first proposed by Sayre.[6] It has properties resembling the Patterson, but from a practical point of view it is widely different from it. Thus its calculation involves (for a noncentrosymmetric crystal) the complete knowledge of the structure factor, that is, both its magnitude and phase. The Patterson, on the other hand, requires only a knowledge of the magnitude of $|F|$, which is known from experiment.

From the relation $F^2 = FF$ it is readily seen that the squared structure is the convolution of the crystal structure with itself. It would thus contain N

peaks of strength f_j^2 at $2\mathbf{r}_j$ and $N(N-1)/2$ peaks of strengths $2f_if_j$ at $(\mathbf{r}_i + \mathbf{r}_j)(i > j)$. If the structure is centrosymmetric, $\tilde{F} \equiv F$, so that $|F|^2 \equiv F^2$. Thus the squared structure and the Patterson structure become identical for a centrosymmetric crystal.

It is worthwhile noting that the coefficients used in the squared structure are related to those of the Patterson by the relation $F^2 = |F|^2 \exp(2i\phi)$. The list of peaks in the Patterson and squared structures are summarized in Table 1.

Modulus structure[4,7]

We have seen earlier that the structure factor $F_N = |F_N| \exp(i\phi_N)$, when fed into a Fourier synthesis, leads to the structure F_N. It is of interest to examine the nature of the Fourier synthesis obtained by feeding in separately the modulus part, $|F_N|$, and the phase factor, $\exp(i\phi_N)$, in a Fourier synthesis. We shall consider the former first since, as will be shown presently, it has properties very similar to the Patterson synthesis. The nature of the synthesis using the phase factor alone will be considered later in Ch. 4.

The properties of the modulus synthesis were first deduced by Ramachandran and Raman[4] and may be shown as follows. Starting from the fundamental relation (41a) of Ch. 1, it is readily seen that

$$|F_N(\mathbf{H})|^2 = \sum_{j=1}^{N} f_j^2 + \sum_{\substack{j \\ j \neq k}} \sum_k f_j f_k \exp[2\pi i \mathbf{H} \cdot (\mathbf{r}_j - \mathbf{r}_k)]. \qquad (10)$$

When fed into a Fourier synthesis, of the type of (44a) of Ch. 1, the first term in the right-hand side leads to a large origin peak of strength S_N^2, as given by (9), and the second term leads to a number of scattered peaks of smaller strengths $f_j f_k$. Writing (10) as

$$|F_N|^2 = S_N^2 \left\{ 1 + \sum_j \sum_k \frac{f_j f_k}{S_N^2} \exp[2\pi i \mathbf{H} \cdot (\mathbf{r}_j - \mathbf{r}_k)] \right\}, \qquad (11)$$

we may write the coefficient $|F_N|$ of the modulus synthesis in the form

$$|F_N| = S_N \left\{ 1 + \sum_j \sum_k \frac{f_j f_k}{S_N^2} \exp[2\pi i \mathbf{H} \cdot (\mathbf{r}_j - \mathbf{r}_k)] \right\}^{1/2}. \qquad (12)$$

Since S_N^2 is, in general, large compared with $f_j f_k$, we may expand the square root as a binomial series, and as a first approximation, take only the first term in the expansion. We thus obtain

$$|F_N| = S_N + \sum_j \sum_k \frac{f_j f_k}{2S_N} \exp[2\pi i \mathbf{H} \cdot (\mathbf{r}_j - \mathbf{r}_k)]. \qquad (13)$$

It is clear from (13) that the principal peaks in the modulus synthesis are exactly at the same positions as in the Patterson synthesis (Table 2). Their strengths are however different. Apart from the changes in the absolute values of the peak strengths, the ratio of the nonorigin to the origin peak is $\frac{1}{2}(f_j f_k/S_N^2)$ in the modulus synthesis, as compared with the value $(f_j f_k/S_N^2)$ in the Patterson. Thus this ratio is only half as much as that of the Patterson synthesis in the case of the modulus synthesis. Essentially, however, the modulus synthesis has the same nature as the Patterson synthesis and it is basically a vector diagram of the original structure.

The syntheses $|F_N|^{2n}$ and F_N^n

In fact the method applied above to examine the nature of the modulus synthesis can be extended to the general case of a synthesis using in general $(|F_N|^2)^n$ as coefficient. It is readily shown[7] that the $(|F_N|^2)^n$ synthesis contains as a first approximation an origin peak of strength S_N^{2n} and principal nonorigin peaks, at the same positions as in the Patterson, namely at $(\mathbf{r}_j - \mathbf{r}_k)$, having strengths $nf_j f_k S_N^{2n-2}$. The ratio of the strength of a nonorigin peak to that of an origin peak is n times that in the ordinary Patterson (Table 2).

Table 2   Peaks[†] in the modulus synthesis and
the synthesis $(|F|^2)^n$

Position	Strength	
	Modulus synthesis	Synthesis $(|F|^2)^n$
Origin	S_N	S_N^{2n}
$(\mathbf{r}_j - \mathbf{r}_k)$	$f_j f_k/2S_N$	$nf_j f_k S_N^{2n-2}$
Ratio of nonorigin to origin peak	$f_j f_k/2S_N^2$	$nf_j f_k/S_N^2$

† First approximation, for the general noncentrosymmetric case.

In those cases in which n is an integer, and not a half integer, however, the structure $|F_N^2|^n$ can be obtained exactly. We may illustrate this by considering $|F_N^2|^2 \equiv |F_N|^2 \cdot |F_N|^2$. By a simple application of the convolution principles discussed at the beginning of this chapter, the peaks are first seen to be of the types

a. S_N^4 at origin.

b. $2f_j f_k S_N^2$ at $\mathbf{r}_j - \mathbf{r}_k$, $(j \neq k)$.

c. $f_j f_k f_l f_m$ at $\mathbf{r}_j - \mathbf{r}_k + \mathbf{r}_l - \mathbf{r}_m$, $(j \neq k, l \neq m)$.

However, there will be a feedback from the peaks of class (c) into those of class (b) by the cases when $j = m$, or $k = l$, which makes the strength at $\mathbf{r}_j - \mathbf{r}_k \simeq 4f_j f_k S_N^2$. Similarly, there is a feedback from the examples $j = m$ and $k = l$, in class (c) which contribute to the strength at the origin, of the order of S_N^2, which is small compared to S_N^4.

The case of F_N^n, where n is an integer, is very simple to obtain. Writing $F_N = \sum f_j \exp(2\pi i \mathbf{H} \cdot \mathbf{r}_j) = \sum F_j$,

$$F_N^n = (\sum F_j)^n = \sum_{j, k, \ldots, p = 1}^{N} F_j F_k \cdots F_p,$$

where all combinations of $j, k, \ldots, p$ with $j, k, \ldots, p$ going from 1 to N are possible. This leads in general to peaks of strength $f_j f_k \cdots f_p$ at $\mathbf{r}_j + \mathbf{r}_k + \cdots + \mathbf{r}_p$. As in the case of the squared synthesis, some of these may superpose and thus lead to double, triple, or other multiple peaks. Only the method of working these out is indicated here and the details are left to the reader.

The properties deduced above may be illustrated with an actual example. Figure 1 shows the syntheses calculated with coefficients $|F|$, $|F|^2$, and $|F|^4$,

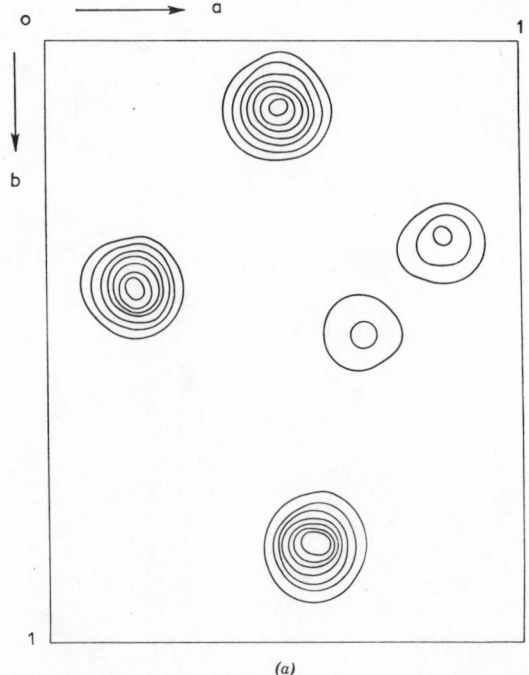

(a)

Fig. 1. *(a)* The Southern Cross taken as a five-atom model used to calculate the syntheses in parts *b*, *c*, and *d*.

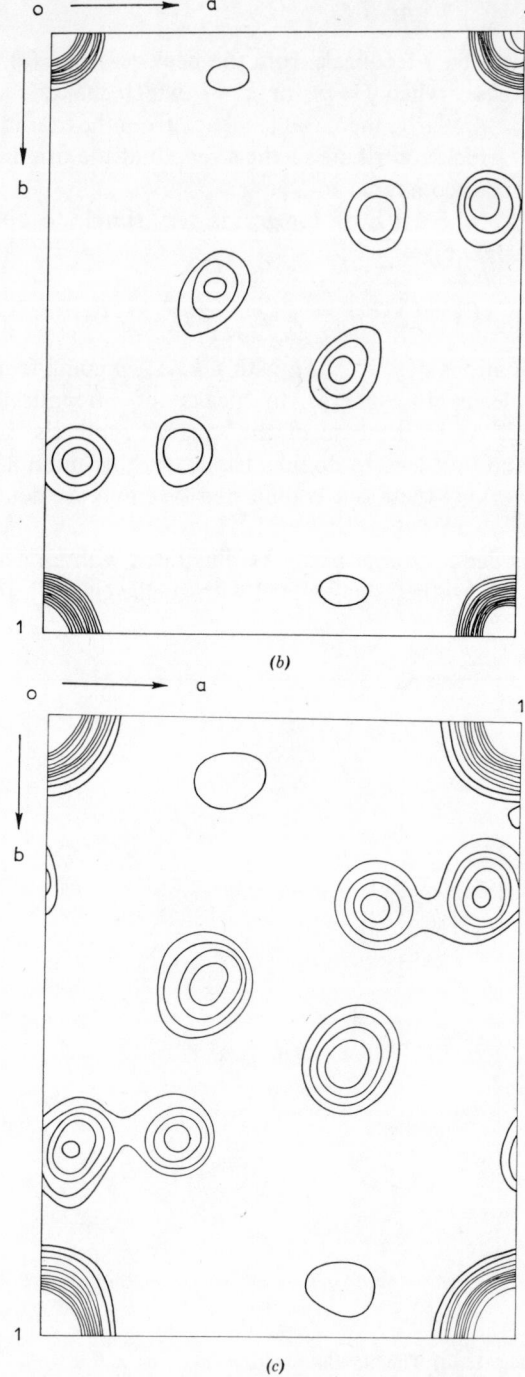

Fig. 1. (*b*) Synthesis using $|F|$; (*c*), synthesis using $|F|^2$.

o   →   a

b

1

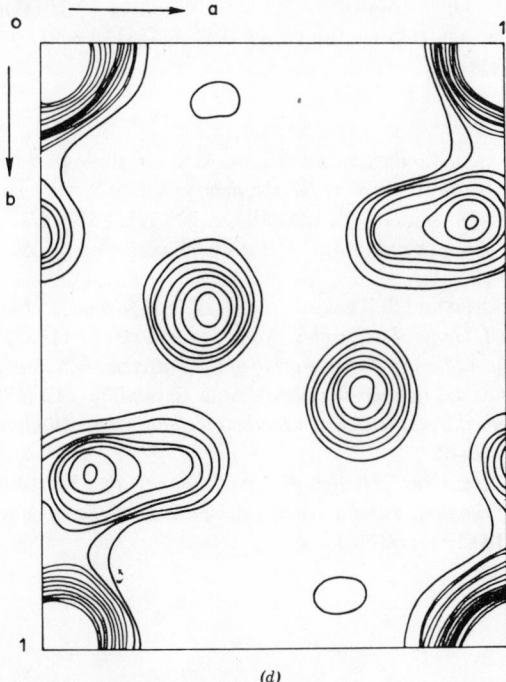

(d)

Fig. 1. (d) synthesis using $|F|^4$ as coefficients. Contours are at arbitrary intervals. Note the similarity of the diagrams (b), (c), and (d) above.

respectively, on a hypothetical structure (Fig. 1a). The model, in noncentrosymmetric plane group p1, was taken to be a five-atom structure resembling the Southern Cross. The strengths of these were taken to be approximately inversely proportional to the magnitudes of the stars. As is to be expected from the earlier discussion, all the three maps, namely $|F|$, $|F|^2$, and $|F|^4$, are quite similar in nature. The average of the ratio of a nonorigin peak to origin peak in the three syntheses are 0.14, 0.23, and 0.4, which is in broad agreement with the theoretical expectation that these should be in the ratio of $1:2:4$.

In view of these, it is to be expected that a synthesis which uses as coefficients a power series in $|F|$ will also exhibit Patterson-like properties. Thus, for example, a modified Patterson-type function was devised by Cowley[1] as an admixture of various orders of terms. He suggested a function of the type

$$G(H) = |F(H)|^2 - 3|F(H)|^4 + 4|F(H)|^6 - |F(H)|^8, \tag{14}$$

which was found to improve the ordinary Patterson function by a slight sharpening effect. The function $G(\mathbf{H})$ approximates to $|F(\mathbf{H})|^2$ in the range $0 \leqslant |F(\mathbf{H})| \leqslant 0.15$, whereas in the range $0.15 \leqslant |F(\mathbf{H})| \leqslant 0.65$ it is approximately linear in $|F(\mathbf{H})|$.

References

[1] J. M. Cowley. *A modified Patterson function*. Acta Crystallogr. **9** (1956) 397–398.

[2] A. L. Patterson. *A direct method for the determination of the components of interatomic distances in crystals*. Z. Kristallogr. **90** (1935) 517–521.

[3] A. L. Patterson. *An alternative interpretation for vector maps*. Acta Crystallogr. **2** (1949) 339–340.

[4] G. N. Ramachandran and S. Raman. *Syntheses for the deconvolution of the Patterson function. Part I. General principles*. Acta Crystallogr. **12** (1959) 957–964.

[5] S. Raman. *Syntheses for the deconvolution of the Patterson function. Part II. Detailed theory for noncentrosymmetric crystal*. Acta Crystallogr. **12** (1959) 964–975.

[6] D. Sayre. *The squaring method. A new method for phase determination*. Acta Crystallogr. **5** (1952) 60–65.

[7] R. Srinivasan. *Syntheses for the deconvolution of the Patterson function. IV. Refinement of the theory and a general comparison of the various syntheses*. Acta Crystallogr. **14** (1961) 607–611.

3

Patterson function and the theory of images

The Patterson as a sum of images

The interpretation of the Patterson function as one representing inter-atomic vectors, discussed in the last chapter, can be extended and developed further in a systematic way which helps us to arrive at a purely geometrical approach to the understanding of the Patterson structure in relation to the original crystal structure. This particular method of approach was first discussed by Wrinch.[33,34] The general principles involved in this approach, particularly in relation to the techniques of crystal structure analysis, were studied systematically and thoroughly by Buerger.[2] These investigations show that, in principle, the structure can be solved for from the Patterson function by certain procedures involving purely geometrical operations and without the necessity of calculating the phase angle. A detailed discussion of these is outside the scope of this monograph.[†] We shall, however, outline some of the main results and procedures of these so-called "image methods," insofar as they have a direct bearing on the Fourier method of approach to be considered in succeeding chapters.

Noncentrosymmetric structure

Consider first a noncentrosymmetric structure containing N point atoms of strengths f_j with position vectors $\mathbf{r}_j (j = 1$ to $N)$. Following the notation introduced in Ch. 2, we may denote any single vector $(\mathbf{r}_j - \mathbf{r}_i)$ by the symbol[†]

[†] A particularly extensive treatment of the subject, along with numerous references to literature, is given in the book by Buerger.[4]

$[ij]$. The set of all vectors from any particular atom, say $i = I$, may be denoted by $\{Ij\}$, where we have used the braces to denote a set of vectors with j going from 1 to N. It is easy to see that the total of N^2 vectors in the Patterson may be looked upon as the set of vectors $\{ij\}$, with i, j both going from 1 to N. If we take the original structure $\mathbf{r}_j$ and draw the vectors from a given atom I to all the other atoms j (j including also I), we get the vectors $\{Ij\}$.‡ If now point atoms of strength f_j are located at the termini of the vectors $\{Ij\}$, we obtain a constellation of atoms whose relative positions are exactly the same as in the original structure, except that the atom I is taken to be the origin. This constellation may be called the image of the structure as seen from atom I, or the image I, which may be represented symbolically by $\{Ij\}$. It is now obvious that the Patterson may be considered as a superposition of N such images, each image being seen from one of the N atoms in the original structure. The original structure is therefore embedded in the Patterson. The aim, in all the vector methods, is to recover any one of the images, making use of the geometrical relationships between them.

Suppose now that we make two copies of the Patterson diagram and displace one of them relative to the other by an interatomic vector, say [12], so that the origin of one is placed on the peak [12] of the other unshifted diagram. We may call the unshifted diagram as $\mathscr{P}_0$ and the shifted diagram as $\mathscr{P}_{12}$. It can be shown that a set of peaks in the two diagrams coincide and each pair of coincident peaks is governed by an equation of the type

$$[ij] + [jk] = [ik]. \qquad (1)$$

Thus, for a given vector shift [12], the coincident peaks are given by

$$\{2i\} + [12] = \{1i\}, \qquad (2)$$

$$\{j1\} + [12] = \{j2\}. \qquad (3)$$

The above relations imply that the vector shift [12] produces coincidences of the sets of peaks $\{1i\}$ and $\{j2\}$ of the unshifted map with the sets $\{2i\}$ and $\{j1\}$ of the other shifted map. The coincident peaks in either one of the diagrams consist of two images of the structure. Taking those in the unshifted map, namely, $\{1i\}$ and $\{j2\}$, these correspond to the structure as seen from atom 1 and the inverse of the structure as seen from atom 2. Let us denote

---

† The treatment here follows closely that of Ramachandran.[17,18]

‡ As per the original terminology first introduced by Wrinch[33,34], the set $\{Ij\}$, with $j \neq I$, is termed the image of the $(N-1)$ points in the point I. More generally the set $\{Ij\}$ with $j = 1$ to P is termed the image of the polygon P in the point I. Obviously one could take $j = 1$ to N to include all the N points ($P = N$), including $j = I$ (see Buerger,[4] p. 185). We use in this monograph the term image to include $j = I$ also.

these by the symbols $\mathscr{I}_1$ and $\mathscr{I}'_2$, the prime denoting an inverse image. Symbolically the above result may be represented by

$$\mathscr{P}_0 \times \mathscr{P}_{12} = \{1i\} + \{j2\} = \mathscr{I}_1 + \mathscr{I}'_2. \tag{4}$$

The multiplication symbol $\times$ in $\mathscr{P}_0 \times \mathscr{P}_{12}$ stands for the choice of points common to the diagram $\mathscr{P}_0$ and $\mathscr{P}_{12}$, and the symbol $+$ in $\mathscr{I}_1 + \mathscr{I}'_2$ represents the totality of points contained in both $\mathscr{I}_1$ and $\mathscr{I}'_2$. This use of the symbols $+$ and $\times$, for the "union" and "intersection" of two sets, is similar to that adopted in set theory.

Thus the result of a single vector shift on the Patterson of a noncentrosymmetric structure and the selection of coincident points leads not completely to the structure, but to two images, one of the structure and the other of its inverse. It is easy to see that, to recover the structure completely, one should make another vector shift through another vector $[1j]$, where j is different from 2. Suppose it is to the point $[13]$. This superposition would obviously give $\mathscr{I}_1 + \mathscr{I}'_3$. The images common to the two superpositions with vector shifts $[12]$ and $[13]$ will therefore contain the image $\mathscr{I}_1$ alone, since the images $\mathscr{I}'_2$ and $\mathscr{I}'_3$ will not, in general, have common coincident peaks. This may be represented symbolically by the equation

$$(\mathscr{I}_1 + \mathscr{I}'_2) \times (\mathscr{I}_1 + \mathscr{I}'_3) = \mathscr{I}_1. \tag{5}$$

It may be noticed that the two points in the Patterson to which vector shifts should be made to recover the structure should be such that they arise from interactions with one atom in common. The above procedure can be extended further. If successive vector shifts of the type $[12]$, $[13]$, $[14]$, ... $[1N]$ are made, the common superposition of all of these will again lead to the structure $\mathscr{I}_1$. This is the method of N-fold superposition which has some practical advantages in actual crystal structures, as will become clear later.

The principles outlined above are the bases of the vector-shift methods originally introduced by Clastre and Gay,[5] Garrido,[6] and by Beevers and Robertson.[1] An analytical approach essentially involving the same principles was given by McLachlan.[12,13]

Centrosymmetric case

We shall now consider the centrosymmetric case. As we have seen in Ch. 2, a single peak in the centrosymmetric case is of type $[jj']$, while a double peak is of the type $[ij']$ or $[j'i]$, where $i, j = 1$ to $M(=N/2)$, and a prime indicates an atom in the inverse structure. If we make a vector shift to a single peak, say $[11']$, then the coincidences are governed by the relations

$$\{i1\} + [11'] = \{i1'\}, \tag{6}$$

$$\{1'j\} + [11'] = \{1j\}. \tag{7}$$

Thus we have

$$\mathscr{P}_0 \times \mathscr{P}_{11'} = \{i1'\} + \{1j\} = \mathscr{I}_1 + \mathscr{I}'_1. \tag{8}$$

However, for a centrosymmetric case, we also have the result

$$\{1i\} = \{i1'\} \quad \text{or} \quad \mathscr{I}_1 \equiv \mathscr{I}'_1. \tag{9}$$

so that the two images are identical. Hence we get only the structure and no other points in the superposed set. This leads to the interesting conclusion that, for a centrosymmetric structure, a single vector shift (provided it is to a point corresponding to a single peak) will yield the structure without any duplication. It may be verified readily that a vector shift to a double peak, say of the type [12], does not lead to the structure uniquely, since the two images $\mathscr{I}_1$ and $\mathscr{I}'_2$ are not identical. However, it may be pointed out that, because of the following internal relation between the vectors, namely $\mathscr{I}'_2 \equiv \mathscr{I}_{2'}$, the two images $\mathscr{I}_1$ and $\mathscr{I}'_2$ may be considered to be the same identical direct image $\mathscr{I}$ as seen from points 1 and 2', respectively. Thus the two images obtained are simply related by a translation vector $\mathbf{r}_1 - \mathbf{r}_2$. It may be noticed that in the noncentrosymmetric case this is not true and the two images are not superposable by translation, but only by inversion. The translation relation in the centrosymmetric case can be made use of as will be discussed later in this chapter.

Image-seeking functions

We have been considering so far the case of point atoms and its Patterson. The purely geometrical operations discussed above would lead to the recovery of the structure in such a case. But it is well known that, in actual practice, the atoms have finite size and have a continuous electron-density distribution. The Patterson function for such an actual structure therefore rarely consists of well-resolved peaks. The difficulty is even more acute with projections. Techniques are available for sharpening the peaks by artificial means (as we discussed briefly in Ch. 1) by suitably modifying the coefficients used in the Fourier syntheses. These, however, are invariably accompanied by spurious effects and false details. Although some advantages may be gained by these sharpening methods, the problem still remains of having to deal with the case of non-point-atom vector diagrams. An additional fact to be reckoned with is that the atoms in an actual structure are, in general, of different types and therefore the simple point-atom case has to be modified to take into account the different weights to be attached to the points. A special consideration of these aspects is therefore necessary while applying the theory to actual structures.

This has resulted in the evolution of what are known as image-seeking functions. The basic principle that governs these methods is that a real peak in the Patterson function has the property that, when a vector shift is made to that point, the superposition will reach a maximum. One can have different types of functions to represent quantitatively the amount of superpositions. The three types of functions that are commonly available for this purpose are[2,4]: the sum function, the product function, and the minimum function. These are discussed in great detail in the book by Buerger.[4] We shall briefly consider the principles of these for the case of the superposition of a pair of Patterson maps. Extension to examples of multiple superposition are only indicated.

Sum function. The sum function, as the name implies, takes the sum of the values of the shifted and the unshifted Patterson functions at each point in one of them (say the unshifted diagram) and plots it on another map relative to a suitable origin (say, that of the unshifted Patterson). This map obviously will contain maxima at the regions where the peaks in the two diagrams overlap. However, since it is a sum, all other peaks in the individual Patterson maps will also occur in the map, although they will be of lower strength compared to the superposed peaks. This introduces a certain amount of background containing false detail and makes the interpretation rather difficult.

Product function. The product function utilizes the product of the two superposed values at each point and it can therefore be expected to be better than the sum function in revealing the superposed peaks, since the superposed peaks will have higher strengths in the resultant map relative to the non-superposed ones. Still the background peaks do occur in this case also, as is obvious from the nature of the function.

Minimum function. To obviate the difficulty of background in the product and sum functions, Buerger suggested the minimum function. In this, the minimum of the two values of the superposed maps is taken at every point. As is obvious, the spurious background details will be suppressed in this case and, where peaks overlap, a positive maximum is retained in the minimum function. This function is therefore expected to be the most satisfactory one for the analysis of superposition of Patterson diagrams.

Resolution of images

In all these image-seeking functions we have discussed so far, we have not taken into account the fact that the atoms in the structure could be of different scattering power. So far as the positions of the peaks in the Patterson (and hence the structure extracted from it) are concerned, they are obviously not affected by the differences in scattering power. However, this factor becomes important when we attempt to establish quantitative relations between the image functions and the actual electron density. The image function of the point–atom case can be readily extended by attaching weights to each point in the original structure. In the case of actual atoms, these weights can be taken to be represented by their atomic scattering factors f_j.

Noncentrosymmetric case

Each peak in the Patterson is of strength $f_i f_j$, where f_i, f_j are in general different. Any vector shift to a point, say [12], results in two images, and from (2) and (3), we see that these arise out of coincidences of $2N$ peaks [$2(N-1)$ peaks, if we exclude the pair involving the origin superpositions]. Half of these arise from (2) and the remaining half from (3). It is readily seen that the coincidences involved at the two sets of image points $\{1i\}$ and $\{j2\}$ are symmetrical. Thus in (2) the two images involved are $\{2i\}$ and $\{1i\}$, while in (3), they are $\{j1\}$ and $\{j2\}$. The strength at the superposed point [$1I$] in the unshifted diagram will be $f_1 f_I + f_2 f_I$, say, in the sum function, while at the [$I2$] it will be $f_I f_2 + f_I f_1$, which is also the same. Also, in the product, or minimum function, or any other function that is used, the value of the function at [$1I$] and [$I2$] will be identical, being equal to $\phi(f_1 f_I, f_2 f_I)$, where ϕ is the relevant function. Therefore, if we use any of the image-seeking functions such as the sum, product or minimum function, the resultant pair of images will have identically the same weight. Thus the fact that atoms have different weights does not alter the earlier conclusion that, in general, the structure is duplicated by a center of inversion, when a superposition map is made as a result of a single vector shift. Thus normally there will be no way of distinguishing between the two images.

It is therefore necessary to use multiple vector shifts to solve for the structure in a noncentrosymmetric crystal. In fact, the methods that are now presently used[4,15,29] for this purpose are essentially based on using more than a single vector shift. In this procedure, improvements can be effected, based on two general types of information. The first is the fact that the structure may possess symmetry elements and therefore special information will become available in the form of Harker sections and lines. This may be used in conjunction with a second kind of information (if available), namely, the presence of heavy atoms in the structure, whose interactions (vectors between them) will have a large strength. Therefore peaks at the termini of such vectors, namely the heavy atom–heavy atom interactions, are clearly distinguishable. The presence of symmetry elements also implies the possibility of extracting the minimum function of two minimum functions and still higher-order minimum functions, as has been discussed by Buerger.[4]

The procedures adopted by different workers would appear to differ in the matter of details, though the essential principle involved in all of them is to make the best use of available knowledge regarding symmetry and other partial information about the structure. Thus, in the proposal of Simpson et al.,[29] the symmetry minimum function, which is based upon all the symmetry interactions (e.g., Harker sections and Harker lines), is first set up. A simultaneous higher-order superposition function, which makes use of all symmetry elements in the space group, is also calculated, wherein the symmetry minimum function itself can be used to minimize the number of spurious peaks.

Weighted superposition functions

We had seen above that two images, direct and inverted, which are simultaneously obtained by a single vector shift, are equal in strength if no weighting function is used. We shall now show that one of them can be made relatively stronger, if the shifted Patterson is suitably weighted, in a case in which there are unequal atoms in the structure.[9] To simplify the discussion, suppose the two atoms 1 and 2, between which the vector shift [12] is made, are unequal in strength, e.g., let $f_1 = f_H$ and $f_2 = f_L$, the symbols H and L indicating a heavy and a light atom, respectively, with $f_1(f_H) > f_2(f_L)$. Suppose we take the superpositions involving any one atom I in (2) and (3), which lead to the equations

$$[2I] + [12] = [1I] \quad \text{in} \quad \mathscr{I}_1, \tag{10a}$$

and

$$[I1] + [12] = [I2] \quad \text{in} \quad \mathscr{I}'_2. \tag{10b}$$

If there is no weighting, the values of the minimum function at the points $[1I]$ and $[I2]$ are equal, by the following relations:

$$\text{Min}([2I], [1I]) = f_I \, \text{Min}(f_2, f_1) = f_I f_2, \tag{11a}$$

$$\text{Min}([I1], [I2]) = f_I \, \text{Min}(f_1, f_2) = f_I f_2. \tag{11b}$$

Here, Min (x, y) stands for the smaller of the two quantities x and y. Consider, however, the minimum function of the unshifted Patterson $\mathscr{P}_0$ and the weighted shifted Patterson $W\mathscr{P}_{12}$, where $W = f_1/f_2$. Then

$$\text{Min}\left(\frac{f_1}{f_2}[2I], [1I]\right) = f_I \, \text{Min}(f_1, f_1) = f_I f_1, \quad \text{in} \quad \mathscr{I}_1 \tag{12a}$$

$$\text{Min}\left(\frac{f_1}{f_2}[I1], [I2]\right) = f_I \, \text{Min}\left(\frac{f_1^2}{f_2}, f_2\right) = f_I f_2, \quad \text{in} \quad \mathscr{I}'_2. \tag{12b}$$

Hence each peak in $\mathscr{I}_1$ is stronger than the corresponding peak in the inverse image by the ratio f_H/f_L, and if this ratio is large, the correct image will stand out against a weak inverse image. The method, however, requires for its success the possibility of picking out a peak at the terminus of the vector [12] (i.e., between a heavy atom and a light atom) for making the vector shift.

It is of interest to note that the weighting procedure mentioned above does not work for the product function and also that it is not highly discriminatory with the sum function. It works best with the minimum function.

Such weighting procedures could also be used to obtain higher-order minimum functions using multiple superpositions.[4,8,29] It is even possible to relate theoretically the final image that is extracted to the actual electron density.

Centrosymmetric case

The features discussed earlier for point atoms can be carried over to the case of finite atoms, even when the atoms are not all alike. This is because the vector shift considered was due to a single peak, which is an interaction between like atoms. A solution of the structure is therefore possible from the Patterson provided the single peak can be chosen correctly. However, the detection of a single peak in the Patterson, in the midst of the double peaks, which are much larger in number, is not easy, except when one or two of the atoms happen to be relatively much stronger than the rest. Presence of symmetry elements in the structure may be of help, as in the noncentrosymmetric case, in checking such a choice. However, there is no guarantee that such an approach will be successful especially when the atoms are more or less alike.

Considering that a single peak has been located, any one of the image-seeking functions can be used. If the structure contains a heavy atom of sufficiently large strength, then any of the Fourier methods suggested in Ch. 5, 6, and 7 could be used. The purely geometrical method, however, has an advantage, especially when the heavy atoms are not large enough to dominate the signs of a majority of reflection, for the image function, applied on a single peak, contains in principle the structure and is independent of the strength of the single peak chosen. This prompted some workers[26-28] to suggest that the image function so obtained can be Fourier inverted to obtain the signs of the reflections. The minimum function being the best out of the image-seeking functions, the Fourier inversion is done with this function[26] to yield an approximation to $F(\mathbf{H})$ in the form

$$F(\mathbf{H}) = V \int M(\mathbf{r}) \exp(2\pi i \mathbf{H} \cdot \mathbf{r}) \, dv_r, \tag{13}$$

where $M(\mathbf{r})$ is the electron density of the minimum function. The possibility of calculating such functions in practice depends on analytical procedures and the availability of computers, since the function $M(\mathbf{r})$, in general, is a continuous one, but not in the form of the sum of electron distributions of atoms of known strengths and sizes. Both these difficulties are fortunately not insurmountable, for the minimum function has an equivalent analytical representation, as shown later in this chapter.

If the peak chosen for the vector shift is not a single peak, but say a double peak, it is easily shown that the structure is not obtained uniquely, but that there is a duplication of the image. However, as we have seen earlier, the duplication of the image in this case, unlike in the noncentrosymmetric case, is by a parallel-shift vector. If the peak chosen is a multiple peak, the images obtained are related, in general, by parallel-shift vectors. The number of images depends on the multiplicity of the peak and, in principle, if these vectors could be determined the structure could be solved. Thus, although one single shift to a multiple peak cannot produce the complete structure (in which respect the situation is similar to the case of a noncentrosymmetric structure) the information that the multiple images are related by parallel vectors can be used to advantage. For instance, the method of Fourier inversion of the image function may be helpful in this regard. Thus, the Fourier coefficients of the minimum function are calculable [by using (13) or a suitably modified analytical sampling procedure[7]] for the multiple-peak superposition. A Patterson of this, calculated using the squares of the coefficients, gives peaks which provide information regarding the vectors by which the multiple images are shifted in the original minimum function. A higher-order minimum function, using the knowledge regarding these vector shifts, gives a good approximation to the actual structure.

Representation of the image-seeking function as a Fourier synthesis

Our main interest in these image-seeking functions is from the point of view of possible Fourier syntheses by which they can be represented. The methods described earlier are themselves of direct applicability in structure determination, since they involve geometrical operations and simple algebraic manipulations which are easily handled, especially if a computer is available. Nevertheless, the possibility of representing some of them as Fourier syntheses is of much interest, both from theoretical as well as practical points of view.

The sum function

The first such analytical approach to the superposition function was given by McLachlan,[12,13] who represented the sum function as a mixed series of Fourier syntheses. The method can be understood by first considering the effect of a single vector shift. Suppose we consider a peak [12] in the Patterson and form the sum function of the two maps corresponding to this vector shift.

The unshifted Patterson is represented by the Fourier series using $|F|^2$ as coefficients. The Patterson shifted by $[12] = \mathbf{r}_2 - \mathbf{r}_1$ is equivalent, by (28) of Ch. 1, to a Fourier synthesis using the coefficients $|F|^2 \exp\left[-2\pi i\mathbf{H} \cdot (\mathbf{r}_2 - \mathbf{r}_1)\right]$. This is so because a translation leads simply to multiplication by a phase factor. The sum function is therefore equivalent to performing a Fourier synthesis using $|F|^2\{1 + \exp\left[-2\pi i\mathbf{H} \cdot (\mathbf{r}_2 - \mathbf{r}_1)\right]\}$. The above process can be extended further. If we have a set of shift vectors $(\mathbf{r}_j - \mathbf{r}_1)$, $j = 1$ to P, (i.e., we allow the origin of the shifted Patterson to rove over the termini of these P vectors), then the sum function for all the vectors $j = 1$ to P is the Fourier transform of

$$|F|^2 \left\{ 1 + \sum_{j=2}^{P} \exp\left[-2\pi i\mathbf{H} \cdot (\mathbf{r}_j - \mathbf{r}_1)\right] \right\}$$

$$= |F|^2 \left\{ \sum_{j=1}^{P} \exp\left(-2\pi i\mathbf{H} \cdot \mathbf{r}_j\right) \right\} \exp\left(2\pi i\mathbf{H} \cdot \mathbf{r}_1\right). \tag{14}$$

This is the basis of the vector convergence method of Beevers and Robertson.[1] The last factor in (14) only represents a phase factor and is equivalent to translating the required image by its vector $\mathbf{r}_1$. A modified weighted form of this sum function leads to the alpha synthesis, which is discussed in Ch. 5.

The product function

In this case, since we require the product of the values at the ends of a vector, say [12], it is equivalent to taking the Fourier coefficients to be equal to

$$|F(\mathbf{H})|^2 \sum_{\mathbf{H'}} |F(\mathbf{H'})|^2 \exp\left[-2\pi i\mathbf{H'} \cdot (\mathbf{r}_2 - \mathbf{r}_1)\right],$$

and performing the synthesis

$$\sum_{\mathbf{H}} \sum_{\mathbf{H'}} |F(\mathbf{H})|^2 |F(\mathbf{H'})|^2 \exp\left[-2\pi i(\mathbf{H} + \mathbf{H'}) \cdot \mathbf{r}_2\right] \exp\left(2\pi i\mathbf{H} \cdot \mathbf{r}_1\right). \tag{15}$$

This can be extended to higher-fold superpositions, but the formulas are clearly not simple.

The minimum function

The Fourier representation of the minimum function was given by Taylor[31] (see also Buerger[3]). This, however, differs from the other two image-seeking functions discussed earlier in that it cannot be calculated as a single Fourier with coefficients fed in conveniently, but involves taking the modulus of a quantity which can be either positive or negative. Such a process can, however, be handled readily by a computer because it is a simple logical operation.

Taylor's result[31] is based on the fact that, given two quantities a and b, the minimum of the two, $M(a, b)$, can be represented by the equation

$$M(a, b) = \tfrac{1}{2}[(a + b) - |a - b|].$$ (16)

We may choose in the above a and b to represent the values of the Patterson function $\mathscr{P}_0$ and $\mathscr{P}_{12}$ at any point $\mathbf{r}$. Thus, denoting by $\mathscr{P}(\mathbf{r})$ the value of the Patterson function at the position $\mathbf{r}$ and by $\mathbf{t}$ the shift vector $[12] = \mathbf{r}_2 - \mathbf{r}_1$, or a general shift vector, as necessary, the minimum function $M_t(\mathbf{r})$ at $\mathbf{r}$, corresponding to the shift $\mathbf{t}$, takes the form

$$M_t(\mathbf{r}) = M[P(\mathbf{r}), P(\mathbf{r} - \mathbf{t})].$$ (17)

If, for convenience, we choose the origin of the superposition diagram at the midpoint of the shift vector $\mathbf{t}$, which is a center of inversion, we have

$$M_t(\mathbf{r}) = M\left[P\left(\mathbf{r} + \frac{\mathbf{t}}{2}\right), P\left(\mathbf{r} - \frac{\mathbf{t}}{2}\right) \right]$$ (18)

$$= S(\mathbf{r}) - |D(\mathbf{r})|$$ (19)

where

$$S(\mathbf{r}) = \tfrac{1}{2}\left[P\left(\mathbf{r} + \frac{\mathbf{t}}{2}\right) + P\left(\mathbf{r} - \frac{\mathbf{t}}{2}\right) \right]$$ (20a)

and

$$D(\mathbf{r}) = \tfrac{1}{2}\left[P\left(\mathbf{r} + \frac{\mathbf{t}}{2}\right) - P\left(\mathbf{r} - \frac{\mathbf{t}}{2}\right) \right].$$ (20b)

The Fourier series for S and D are readily found by substituting the expression

$$P(\mathbf{r}) = \sum_{\mathbf{H}} |F(\mathbf{H})|^2 \cos (2\pi\mathbf{H} \cdot \mathbf{r})$$ (21)

in (20a) and (20b). We thus obtain

$$S(\mathbf{r}) = \sum_{\mathbf{H}} |F(\mathbf{H})|^2 \cos (2\pi\mathbf{H}\cdot\mathbf{r}) \cos (2\pi\mathbf{H}\cdot\mathbf{t}),$$ (22a)

$$D(\mathbf{r}) = \sum_{\mathbf{H}} |F(\mathbf{H})|^2 \sin (2\pi\mathbf{H}\cdot\mathbf{r}) \sin (2\pi\mathbf{H}\cdot\mathbf{t}).$$ (22b)

Thus it is seen that the minimum function can also be represented as a Fourier series, although its calculation in practice differs from the other two.

It may be mentioned that another analytical method has been suggested by McLachlan[14] for the operation of choosing the minimum of two quantities. Thus, if we have two quantities a and b, and the smaller of the two is to be obtained, we could take the sum of the reciprocals of a large nth power of the two quantities. Thus

$$\frac{1}{a^n} + \frac{1}{b^n} \simeq \frac{1}{b^n}, \quad \text{if} \quad b < a, \quad n \text{ large,}$$

and as a first approximation, the nth root of $(1/a^n + 1/b^n)$ gives the smaller of the two quantities. This may be useful in handling the minimum function on a computer, but it appears that (16) is simpler.

Relation between image functions and electron densities

The relation between the final values of the "image-function" obtained by any of the image-seeking functions and the actual electron densities expected in a Fourier synthesis can be given in the form of inequalities.[2,4,8] We shall not go into the details of these, as they are treated extensively by Buerger.[4]

Cumulative functions[16,17,21,22]

It is useful, in connection with superposition functions, to introduce what might be called the "cumulative function." Thus, given a Patterson or Patterson-type function, any vector shift $t = r_2 - r_1$, which corresponds to a real interatomic vector in the structure, leads, on application of any of the superposition functions, to the recovery of the structure and its inverse. We shall call the integral of the image (superposition) function over the unit cell as the cumulative functions $C(t)$ corresponding to the shift vector t:

$$C(t) = \int \phi[P_0(r), P_t(r)] \, dv_r, \tag{23}$$

where ϕ stands for the operation of one of the image-seeking functions. It can be verified that, when ϕ is the sum function, the integral is a constant for all values of t (equal to twice the integral of $P(r)$ over the unit cell) and does not show any variation from point to point. On the other hand, $C(t)$ varies with t if ϕ is chosen to be the product, or minimum function. In these cases, the value of $C(t)$ for any vector t is a quantitative measure of the extent of overlap of the two functions $P_0(r)$ and $P_t(r)$ for the particular shift vector t. In practice, the integral (23) may be calculated as a sum over the grid points for which $P(r)$ is computed, for t also roaming over these grid points. $C(t)$ is then plotted on a map very similar to the function $P(r)$. We may call the function $C(t)$ also as the "integral" ϕ function. It is obvious that the function $C(t)$ will contain maxima corresponding to vector shifts $t = [ij] = r_j - r_i$, where $[ij]$ corresponds to a peak in the Patterson. However, in general, $C(t)$ is not expected to be an improvement over the ordinary Patterson.

Higher-order cumulative functions can be similarly formulated.[19] Thus, for example, similar to the multiple superpositions discussed earlier, we may define a corresponding cumulative function which is the integral (or sum) of all the image functions for the specified vectors. In principle, corresponding to any operation on the actual values of the Patterson function, we can define an integral of the image function. It is clear, however, that although these are of theoretical interest their possible applications in practice become more and more complicated because of the enormous amount of calculations involved. Fourier representation for the product cumulative function is possible and is given by Raman and Katz.[19]

We might, however, point out a few possible applications of the simple first-order cumulative function $C(t)$. Suppose that the function $P(r)$ is not the usual Patterson function, but a modified one. We shall show in Ch. 10 that, if we calculate $C(t)$ on the difference-Patterson function of a pair of isomorphous crystals, it will exhibit maxima of much higher strength corresponding to a replaceable atom–replaceable atom vector, than the peak in the Patterson itself, but not for other shift vectors.

The possible use of $C(t)$ for distinguishing single peaks from double or multiple peaks of a centrosymmetric structure,[30] does not appear to be promising, but may prove a useful adjunct to other methods. The function $C(t)$ is, in general, useful in various situations in which superposition operations are performed on a vector map, such as the Patterson function or one related to it, in an exploration to find the maximum overlap of images. For instance, one of the possible approaches in protein crystallography is to use the presence of noncrystallographic symmetry (such as a twofold axis) in the molecule. The superposition operation in such a case requires a rotation (in addition to any translation that may be necessary) to find the maxima. The actual location of the symmetry axis, determined this way, may aid in the development of the structure by phase-determining methods.[10,11,23–25,32]

Another possible use of the cumulative function is in conjunction with sharpening of Patterson syntheses.[19] It was mentioned in Ch. 2 that the use of sharpening functions introduces, in general, series-termination errors, which result in certain undesirable features such as negative ripples. Some of the real peaks in such a sharpened Patterson may thus be swamped out by accident.[20] On the other hand, if we calculate the cumulative function $C(t)$ on such a sharpened Patterson, a peak would still be observable in this map corresponding to such locations. This may be particularly useful in an iterative process of recovering the true peaks in a sharpened Patterson which may be lost by series termination effects.

One particular method of sharpening the Patterson which is used in such an iterative process may be pointed out here. This is based on the idea that a proper sharpening of the peaks in the Patterson (without ill effects) requires an extension of the total data in reciprocal space. It has been suggested[20] that the intensities of reflections, even beyond the limit of actually observed data, can be generated artificially by Fourier inversion of the Patterson function obtained from the original set of intensities. This, coupled with suitable sharpening functions, could lead to an improved approximation to point-atom Pattersons.

References

[1] C. A. Beevers and J. H. Robertson. *Interpretation of the Patterson synthesis.* Acta Crystallogr. **3** (1950) 164.

[2] M. J. Buerger. *Vector sets.* Acta Crystallogr. **3** (1950) 87–97.

[3] M. J. Buerger. *An intersection function and its relation to the minimum function of x-ray crystallography.* Proc. Nat. Acad. Sci. U.S. **39** (1953) 678–680.

[4] M. J. Buerger. *Vector space and its application in crystal-structure analysis.* (John Wiley and Sons, 1959).

[5] Jose Clastre and Robert Gay. *La determination des structures cristallines a partir du diagramme de Patterson.* Compt. rend. **230** (1950) 1876–1877.

[6] Jules Garrido. *Sur la determination des structures cristallines an moyen de la transformee de Patterson.* Compt. rend. **230** (1950) 1878–1879.

[7] G. Germain and M. M. Woolfson. *Some ideas on the deconvolution of the Patterson function.* Acta Crystallogr. **21** (1966) 845–848.

[8] von Erwin Hellner. *Zur structur bestimmung mit hilfe von superpositions methoden.* Z. Kristallogr. **108** (1956) 64–81.

[9] R. A. Jacobson and L. J. Guggenberger. *Weighting factors in image seeking methods.* Acta Crystallogr. **20** (1966) 592–593.

[10] P. Main. *Phase determination using noncrystallographic symmetry.* Acta Crystallogr. **23** (1967) 50–54.

[11] P. Main and M. G. Rossmann. *Relationships among structure factors due to identical molecules in different crystallographic environment.* Acta Crystallogr. **21** (1966) 67–72.

[12] Dan McLachlan, Jr. *The determination of crystal structures from x-ray data without a knowledge of the Fourier coefficients.* Paper presented at the Pennsylvania State College Meeting of the Am. Cryst. Assoc. April 10, 1950.

[13] Dan McLachlan, Jr. *The development of some extensions of Patterson's principles and their application to the solution of crystal structures.* Bull. No. 50. Utah Eng. Exp. Sta., Vol. 41, No. 6 (Dec. 1950) 40 pages.

[14] Dan McLachlan, Jr. *Buerger's minimum function expressed as an equation.* Z. Kristallogr. **115** (1961) 305–306.

[15] A. D. Mighell and R. A. Jacobson. *Analysis of three-dimensional Patterson maps using vector verification.* Acta Crystallogr. **16** (1963) 443–445.

[16] V. Raghupathy Sarma and R. Srinivasan. *Principle of maximum superposition: A method for determining the positions of replaceable atoms in isomorphous crystals.* Acta Crystallogr. **15** (1962) 457–460.

[17] G. N. Ramachandran. *Fourier and vector methods in crystal structure analysis.* Physics Conference and Symposium on Solid State Physics Proceedings, Bangalore (1960) 131–150.

[18] G. N. Ramachandran. *Fourier synthesis for partially known structures.* In Advanced Methods of Crystallography. (Academic Press, London, 1964) 25–65.

[19] S. Raman and J. Lawrence Katz. *Accumulation function in x-ray determination of crystal structures.* Z. Kristallogr. **124** (1967) 26–42.

[20] S. Raman and J. Lawrence Katz. *An analytical method of obtaining sharpened Patterson functions.* Z. Kristallogr. **124** (1967) 43–63.

[21] S. Raman and W. N. Lipscomb. *Two classes of functions for the location of heavy atoms and for solution of crystal structures.* Z. Kristallogr. **116** (1961) 314–327.

[22] S. Raman and W. N. Lipscomb. *The Patterson approach to phase problem.* In crystallography and crystal perfection (Ed.) G. N. Ramachandran (Academic Press, London, 1963) 79–84.

[23] M. G. Rossmann and D. M. Blow. *The detection of subunits within the crystallographic asymmetric unit.* Acta Crystallogr. **15** (1962) 24–31.

[24] M. G. Rossmann and D. M. Blow. *Determination of phases by the conditions of non-crystallographic symmetry.* Acta Crystallogr. **16** (1963) 39–45.

[25] M. G. Rossmann, D. M. Blow, M. M. Harding, and E. Coller. *The relative positions of independent molecules within the same asymmetric unit.* Acta Crystallogr. **17** (1964) 338–342.

[26] V. I. Simonov. *Determination of the phase of structure amplitude by a modification of the minimization function.* Soviet Physics—Doklady **6** (1961) 98–100.

[27] V. I. Simonov and B. M. Shchedrin. *The Fourier integral of the minimum function and the signs of structure amplitudes.* Soviet Physics: Crystallography **6** (1961) 288–297.

[28] V. I. Simonov. *A possible automatic approach to the superposition method of solving crystal structures.* Soviet Physics: Crystallography **10** (1965) 116–120.

[29] Paul G. Simpson, Robert D. Dobrott, and William N. Lipscomb. *The symmetry minimum function: High order image seeking functions in x-ray crystallography.* Acta Crystallogr. **18** (1965) 169–179.

[30] R. Srinivasan. Unpublished.

[31] William J. Taylor. *Fourier representation of Buerger's image seeking minimum function.* J. Appl. Phys. **24** (1953) 662–663.

[32] P. Tollin, P. Main, and M. G. Rossmann. *The symmetry of the rotation function.* Acta Crystallogr. **20** (1966) 404–407.

[33] D. M. Wrinch. *The geometry of discrete vector maps.* Phil. Mag. **27** (1939) 98–122.

[34] D. M. Wrinch. *Vector maps of finite periodic point sets.* Phil. Mag. **27** (1939) 490–507.

4

The phase synthesis and other standard syntheses

The phase synthesis[1-3,5,7]

We have seen earlier that, out of the two parts of the structure factor F, namely $|F_N|$ and $\exp(i\phi_N)$, a Fourier synthesis using the former alone as coefficients has properties similar to the Patterson. It would be of interest to consider what would be the nature of the synthesis using $\exp(i\phi_N)$ as coefficients. We may call such a synthesis as the phase synthesis. Its properties can be worked out in the following manner.[1,3] We have the relation

$$|F_N| \exp(i\phi_N) = F_N. \tag{1}$$

This equation tells us that the convolution of the modulus structure with the phase structure must lead to the original structure F_N. The peaks in the latter are known—they are of strength f_j and occur at $\mathbf{r}_j$. We have also worked out the principal peaks in the modulus structure in Ch. 2. We may use these facts to work out the peaks in the phase synthesis as follows.[2] Referring to (1), we ask—what is the nature of the principal peaks in the synthesis $\exp(i\phi_N)$ which satisfy the condition that, when they are convolved with the principal origin peak of strength S_N in the $|F_N|$ structure, they would lead to f_j at $\mathbf{r}_j$ in the real structure. Since we are using our principle in the reverse direction to deduce the peaks, the process must necessarily be one of successive approximations. Thus we have to examine now whether the peaks deduced above for the phase synthesis are the only ones in the phase structure. Obviously this cannot be so, for the presence of peaks f_j/S_N at $\mathbf{r}_j$ would mean that, by interaction with the other peaks in the modulus structure, they would yield new peaks. Thus we are led to the following results.

60

Phase synthesis $\exp(i\phi_N)$, first approximation	Modulus synthesis, $|F_N|$	Convolution F_N	
f_j/S_N at $\mathbf{r}_j$	S_N at 0	f_j at $\mathbf{r}_j$	(2a)
	$\dfrac{f_k f_l}{2S_N}$ at $\mathbf{r}_k - \mathbf{r}_l$	$\dfrac{f_j f_k f_l}{2S_N^2}$ at $\mathbf{r}_j + \mathbf{r}_k - \mathbf{r}_l$	
	$(k \neq l)$	$(k \neq l)$	(2b)

The second set of peaks of the convolution in the last column of (2b) should not really be present in F_N. This leads us to infer that there should be negative peaks in the phase structure, of such strengths and locations that, when they are convolved with the principal origin peak in the modulus synthesis, they would lead to a set of peaks which would cancel out the second-order peaks in the last column in (2b). It is easy to see that these peaks in the phase structure should be at $\mathbf{r}_j + \mathbf{r}_k - \mathbf{r}_l$, and their strengths should be $-f_j f_k f_l/2S_N^3$. We have neglected here, however, the effect of the second-order peaks at $\mathbf{r}_k - \mathbf{r}_l$ in the modulus synthesis. But it can be readily verified that taking these into account would not materially affect the strength deduced, since the contribution from the origin peak is of a higher order of magnitude than that from the nonorigin peaks of the modulus synthesis. Of course the process can be extended further to work out the strengths more accurately, but we shall confine ourselves in this chapter only to the first-order approximation. Thus, to this order of approximation, the phase synthesis contains peaks as given in Table 1.

Table 1 Positions and strengths of the peaks in the syntheses discussed in this chapter

Synthesis	Position	Strength
Phase synthesis[†] Structure $\exp(i\phi_N)$	$\mathbf{r}_j$ $\mathbf{r}_j + \mathbf{r}_k - \mathbf{r}_l (k \neq l)$	f_j/S_N $-f_j f_k f_l/2S_N^3$
Reciprocal synthesis[†] Structure $1/F_N$	$-\mathbf{r}_j$ $-\mathbf{r}_j + \mathbf{r}_k - \mathbf{r}_l (k \neq l)$	f_j/S_N^2 $-f_j f_k f_l/S_N^4$
Phase squared synthesis[†] Structure $\exp(2i\phi_N)$	$2\mathbf{r}_j$ $\mathbf{r}_j + \mathbf{r}_k$	f_j^2/S_N^2 $2f_j f_k/S_N^2$
Cosine synthesis Structure $|F_N| \cos\phi_N$	$\pm\mathbf{r}_j$	$\tfrac{1}{2}f_j$
Sine synthesis Structure $i|F_N| \sin\phi_N$	$\mathbf{r}_j$ $-\mathbf{r}_j$	$\tfrac{1}{2}f_j$ $-\tfrac{1}{2}f_j$

† The results are to the first approximation.

The important feature to be remembered about the phase synthesis is that the positions and relative strengths at the *principal* peaks in it are the same as those in the actual crystal structure (synthesis F). This means that it is the phases that really determine the structure and the magnitudes of the structure factors play only a subordinate role. This may be looked at from another point of view. The phase synthesis may be considered as a Fourier synthesis with all the amplitudes approximated by unity, while the modulus synthesis can be considered as a Fourier synthesis with all the phase angles approximated by zero (or the phase part exp $(i\phi)$ approximated by unity). Of the two drastic approximations, the former still leads to the structure F_N as a first approximation, while the latter does not, and leads only to the peaks in the Patterson.

A quite different study[7] involving statistical analysis of the phases and amplitudes of the structure factors also point to the same result, namely that the structural information is far more heavily contained in the phases than in the amplitudes.

Importance of phase angles[2,5,7]

Phase synthesis

The foregoing conclusions have been illustrated in a striking manner[2,5] in a series of calculations. Initially the phase synthesis was calculated taking all the amplitudes to be equal to a constant value (instead of all being equal to unity), the correct phases (or signs) were used, and the corresponding Fourier syntheses were performed. However, this was found to suffer from a sudden termination of series at some value of $(\sin \theta)/\lambda$. This termination error could be overcome by introducing a gradual falloff of the scattering factor as a function of $(\sin \theta)/\lambda$, so that it is practically zero at the limit of the observed data in reciprocal space used for the Fourier synthesis. In fact, in order to simulate the case of an actual structure in which atoms are of finite size, the falloff of the structure amplitude was taken to be equal to that for a mean atom in the structure. Thus all reflections having a given value of $\sin \theta$ were assigned the same structure amplitude, equal to $(\sum f_j^2)^{1/2}$ for the particular value of $\sin \theta$. This normalized phase synthesis was tested by Srinivasan[5] and is shown in Figs. 1a and 2a. These may be compared with the corresponding actual Fourier syntheses (using both the correct amplitudes and phases) shown in Fig. 1c and Fig. 2c, respectively. The calculations performed were on the structure of L-tyrosine hydrochloride. It may be seen that the phase synthesis resembles the actual Fourier synthesis remarkably well. Even the relative peak strengths of the various atoms have come out correctly. In particular, the heavy atom, chlorine, has come out prominently with almost the same strength as in the actual Fourier synthesis, in spite of the fact that no information regarding the structure amplitudes was utilized.

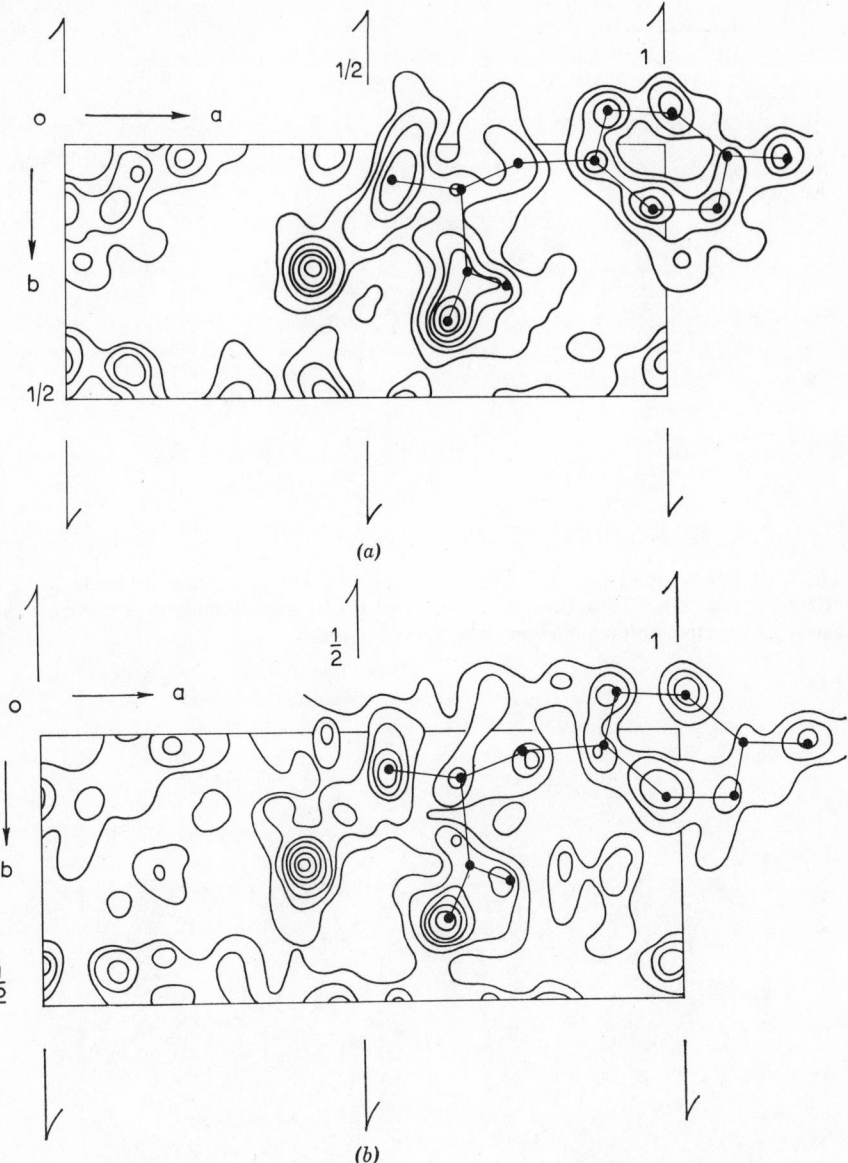

(a)

(b)

Fig. 1 Noncentrosymmetric projection of L-tyrosine HCl. (a) Phase synthesis. The magnitudes of all structure factors are all made equal, except for a decrease with $(\sin \theta)/\lambda$ corresponding to a mean atom in the structure, but the correct phases are used. (b) Random synthesis. The structure amplitudes are randomly permuted among themselves for each range of $(\sin \theta)/\lambda$; the correct phases are, however, correctly used.

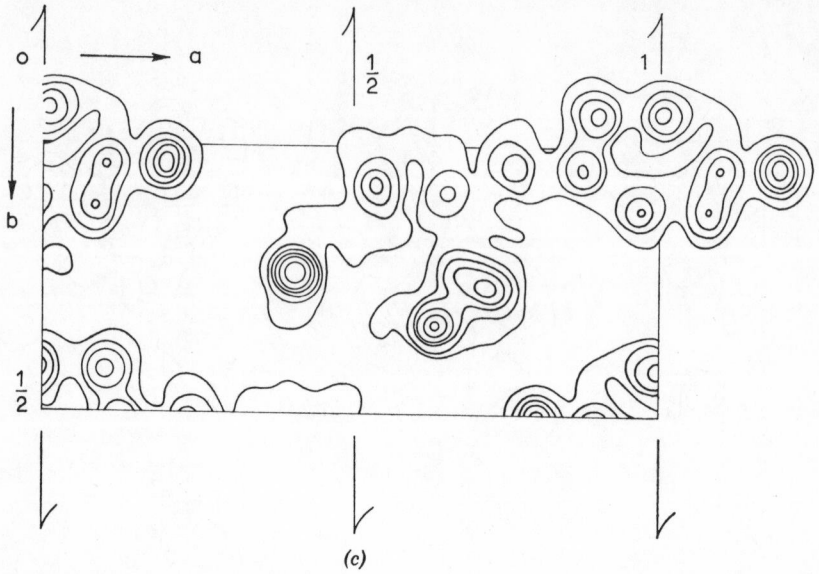

Fig. 1. (*c*) Fourier synthesis. Both correct amplitudes and correct phases are used.

Note the similarity between *a*, *b*, and *c*. In particular the chlorine has come out with much greater strength than the other atoms, even in *a* and *b*.

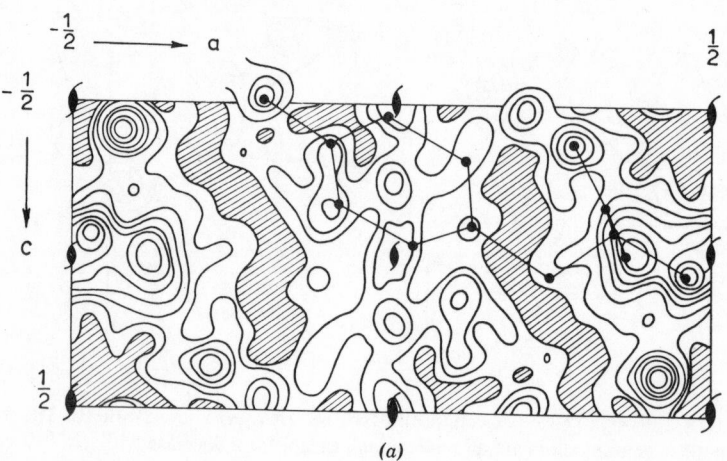

Fig. 2. Centrosymmetric projection of L-tyrosine HCl. (*a*) Phase synthesis. The structure amplitudes are made equal, as in Figure 1*a*, but the correct signs are used.

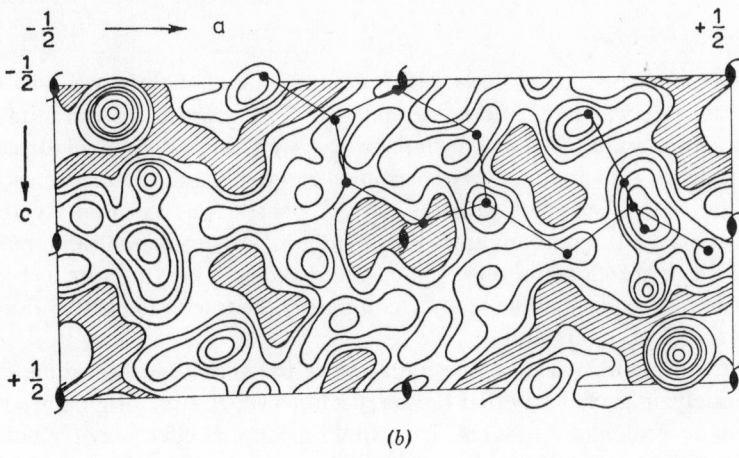

(b)

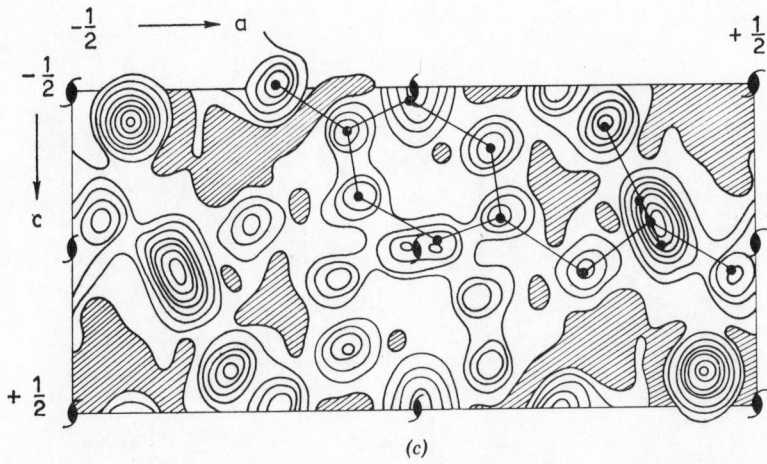

(c)

Fig. 2. (b) Random synthesis. The structure amplitudes are randomly permuted as in Fig. 1b, but the correct signs are used. (c) Fourier synthesis. Both amplitudes and signs are correctly fed in.

The shaded regions have negative electron density. Note again the close similarity between the three diagrams. In particular, the atomic peaks have come out with nearly the correct relative strengths in a and b.

Random synthesis

A few more interesting tests which confirm the above observations in a more concrete way have also been reported.[5] For instance, instead of taking the structure amplitudes as constant for a given sin θ, the actual structure amplitudes in a given range were permuted randomly and these were used in the Fourier synthesis but with the correct phases. The resultant synthesis may be termed the random amplitude synthesis, or simply the random synthesis. These are shown in Figs. 1b and 2b. Here again the structure has come out prominently, as may be seen by comparing them with the actual Fourier syntheses (Figs. 1c and 2c).

In fact, in another test, the permutation of the amplitudes was done more deliberately, in such a way that those reflections which were originally strong were made weak and vice versa. This would mean that effectively the new set of amplitudes were highly anticorrelated (about their mean value) with respect to the original set. Such a synthesis has also been shown[5] to reveal the structure in a fairly good detail although the map was poorer than the phase synthesis and the random synthesis described above.

The overwhelming importance of the phase angles has been demonstrated in another interesting test.[2] Suppose we employ the structure amplitudes $|F_A|$ of one crystal structure A, along with the phases exp $(i\phi_B)$ corresponding

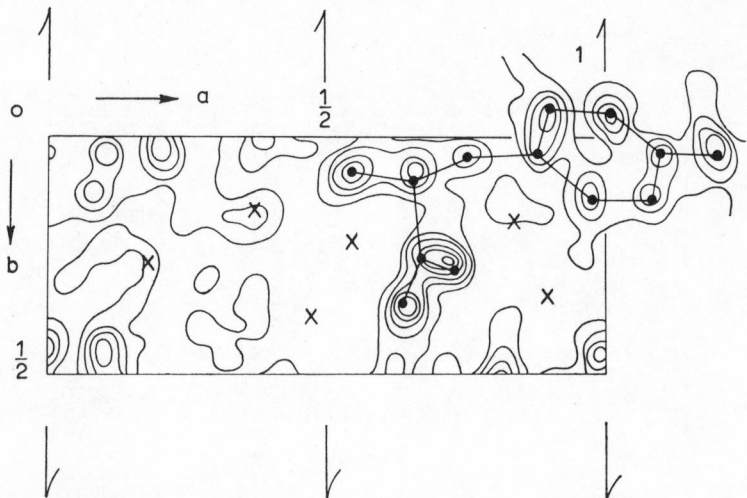

Fig. 3. Fourier synthesis obtained by using the coefficients $|F_A|$ exp $(i\phi_B)$, where $|F_A|$ is the structure amplitude of the structure A (positions marked by crosses) and ϕ_B are the phases of structure B (positions marked by dots). Note the occurrence of peaks only at positions of atoms of structure B, but none at those of structure A.

to an entirely different structure B, and calculate a Fourier synthesis with the coefficients $|F_A| \exp(i\phi_B)$. Such a synthesis is shown in Fig. 3. It again shows peaks only at the positions corresponding to the atoms in the structure B, whose phases were used and none at all at positions corresponding to the atoms of the structure A, whose amplitudes alone were used.

The inescapable conclusion that follows from all these tests is that the phases have an overwhelming importance in revealing the structure and that the amplitudes play only a secondary role in the structure factors. It should be mentioned that this conclusion does not contradict the methods described in Ch. 3, by which the structure is obtained from the Patterson diagram, which can be calculated purely from the amplitudes, without any phase information being put in. However, as was shown in Ch. 2, no Fourier synthesis which uses for its coefficients the intensities $|F(\mathbf{H})|^2$, or any function of these only, yields peaks at atomic positions, but only at locations defined by the interatomic vectors. What is shown here is that, if we require a diagram with peaks at atomic locations obtained by means of a *Fourier synthesis*, the phase information is essential. It so turns out that this information is so overwhelmingly important in this method of approach that reasonably large errors in the amplitudes are tolerated.

Interpretation of results

The reason why the phase synthesis is able to reveal the structure, even under such diverse conditions, can be understood by applying the convolution principles discussed in Ch. 2. All the above syntheses can be interpreted as the convolution of the appropriate modulus synthesis with the phase synthesis $\exp(i\phi_N)$. The latter, as we have seen earlier, contains first-order peaks of strength f_j/S_N at positions $\mathbf{r}_j(j = 1$ to $N)$ corresponding to the structure, from which the phases ϕ_N are derived. The amplitude part used in the random synthesis (or anticorrelated synthesis) does not correspond to any real structure and is not interpretable, in general, in terms of discrete peaks. However, it has one feature in common with the actual modulus synthesis, namely that it also contains an *origin* peak of strength S_N. The peaks in other positions are random and small in strength and they do not have any relation to the corresponding peaks in the modulus synthesis.

A reference to (1) and (2) shows that, in the convolution of the phase part with the amplitude part, it is essentially the origin peak of the modulus synthesis that combines with the first-order peaks of the phase synthesis to yield the structural information f_j at $\mathbf{r}_j$ contained in the standard Fourier. Since the origin peak is also contained in the random synthesis and the anticorrelated synthesis, they would therefore still reveal the structure. In fact, it follows from (A33) that the peak value at the origin of the modulus synthesis is equal to $V^* \sum |F(\mathbf{H})|$, where V^* is the volume of the reciprocal-lattice cell, namely $\mathbf{a}^* \cdot (\mathbf{b}^* \times \mathbf{c}^*)$. Since, in both the random synthesis and the anti-correlated synthesis, the $|F(\mathbf{H})|$'s are only permuted among themselves, we should expect the origin peak to have the same magnitude in these also.

The above arguments are equally applicable to the synthesis $|F_A| \exp(i\phi_B)$ described earlier. What is relevant in this case is only the origin peak of the $|F_A|$ synthesis and the first-order peaks of $\exp(i\phi_B)$. The latter has maxima only at the atomic locations of the structure B, and the combination thus reveals only the atoms corresponding to structure B.

It is interesting to examine this synthesis more closely. We have assumed that the structure A has no relation whatever to the structure B. Thus the modulus synthesis $|F_A|$ will contain an origin peak of strength

$$S_A(S_A^2 = \sum f_{Aj}^2, j = 1 \text{ to } A)$$

and general peaks at $(\mathbf{r}_{Ai} - \mathbf{r}_{Aj})$ of strength $f_{Ai} f_{Aj}/2S_A^2$. These general peaks, at $(\mathbf{r}_{Ai} - \mathbf{r}_{Aj})$, although interpretable in terms of the structure A, have no relation to the peaks of strength f_{Bk}/S_B at $\mathbf{r}_{Bk}$ in the phase structure $\exp(i\phi_B)$. If we now suppose that some of the atoms in A (say $k = 1$ to P) are common to the structure B, we could see clearly that these common peaks, labelled P, of the two structures A and B can interact (namely peaks $f_{Pi} f_{Pj}/2S_A$ at $(\mathbf{r}_{Pi} - \mathbf{r}_{Pj})$ in structure A with peaks f_{Pk}/S_B at $\mathbf{r}_{Pk}$ in structure B) to produce peaks at position $\mathbf{r}_{Pi}$. This leads us to conclude that peaks in structure A could be developed in the synthesis $|F_A| \exp(i\phi_B)$ *only if there are at least a few atoms common to the structures A and B*. Its significance in structure analysis is obvious. In postulating a trial model for a structure, unless the model has at least a few atoms corresponding to the true structure, it is not possible to develop the true structure from the trial structure by the usual Fourier methods. Although this might appear trivial, since it is a fact well known to crystallographers, the above analysis helps us to understand this clearly in terms of convolution theory.

Reciprocal structure[1,3,4,6]

In the remaining sections of this chapter we shall describe other types of synthesis which have properties related to the phase synthesis or use coefficients derived from the phase factor $\exp(i\phi)$.

Let us first consider the reciprocal synthesis, which uses for its coefficients the quantity $1/F_N$. Its properties can be deduced by a method similar to the one adopted earlier for the phase synthesis. Thus, we have the relation

$$\frac{1}{F_N} \cdot |F_N|^2 = \tilde{F}_N, \tag{3}$$

which relates the reciprocal synthesis with the Patterson $|F_N|^2$ and the inverse structure $\tilde{F}_N$, in both of which the peaks are known exactly. Thus one can work out the peaks in $1/F_N$, which are as follows

$$\frac{f_j}{S_N^2} \quad \text{at} \quad -\mathbf{r}_{Nj},$$

$$\frac{-f_j f_k f_l}{S_N^4} \quad \text{at} \quad -\mathbf{r}_{Nj} + \mathbf{r}_{Nk} - \mathbf{r}_{Nl} \quad (k \neq l). \tag{4}$$

Thus it resembles the inverse structure $\tilde{F}_N$ to a first approximation, in the location of the peaks, but with modified peak strengths. This could also have been conjectured in another way, using our knowledge of the phase synthesis. Thus $1/F_N = (1/|F_N|) \exp(-i\phi_N)$; of the two parts, the phase part $\exp(-i\phi_N)$ is nothing but the inverse phase synthesis and should resemble the $\tilde{F}_N$ synthesis. We have also seen earlier that any phase synthesis, in combination with any modulus part, has the overwhelming character of the structure represented by the phase part. Thus we may consider the term $1/|F_N|$ as simply a certain coefficient taken to be the modulus part, which will not affect appreciably the feature of the phase part, namely $\exp(-i\phi_N)$. The net result is that the resultant synthesis has peaks at positions corresponding to the inverse structure as a first approximation.

It follows immediately from the above that the reciprocal inverse structure $1/\tilde{F}_N$ resembles, to a first approximation, the real structure F_N. This is a particularly significant result which we shall use in later chapters.

One aspect of practical importance has to be remembered in connection with the reciprocal synthesis, namely, that in any actual computation that may be attempted, for those reflections for which $|F_N|$ is very small, the corresponding coefficient $(1/|F_N|)$ tends to be very large. Thus the resultant Fourier synthesis will be dominated by such terms only. This is a general feature of syntheses involving such reciprocal terms and one way to overcome this in practice is either to omit these terms completely from the summation, or to use proper weighting terms to suppress their effect. We shall consider such aspects later.

Phase-squared structure

We call the phase-squared structure that synthesis which uses the coefficients $\exp(2i\phi_N)$. Since $\exp(2i\phi_N) = \exp(i\phi_N) \times \exp(i\phi_N)$, this structure is the convolution of the phase structure with itself. The peaks, to a first approximation, are readily worked out to be

$$\frac{f_j^2}{S_N^2} \quad \text{at} \quad 2\mathbf{r}_j,$$

$$\frac{2f_j f_k}{S_N^2} \quad \text{at} \quad \mathbf{r}_j + \mathbf{r}_k. \tag{5}$$

These are similar to the squared structure F_N^2 studied earlier, which is in a way to be expected, since

$$F_N^2 = |F_N|^2 \exp(2i\phi_N),$$

which equation also leads to the same results as (5), to a first approximation.

Cosine and sine syntheses

We shall study now the properties of two types of syntheses which use as Fourier coefficients the real and imaginary parts of the phase factor separately along with the structure amplitudes $|F_N|$. Thus, consider the synthesis $|F_N| \cos \phi_N$, which may be written in the form

$$|F_N| \cos \phi_N = [\tfrac{1}{2}|F_N| \exp(i\phi_N) + \tfrac{1}{2}|F_N| \exp(-i\phi_N)]. \tag{6}$$

The first term has peaks at the same positions as in the correct structure but with half the strength—that is, there are peaks of strength $\tfrac{1}{2}f_j$, at $\mathbf{r}_j$. The second term similarly contains peaks of strength $\tfrac{1}{2}f_j$ at $-\mathbf{r}_j$. The cosine synthesis is the sum of the above two and must, therefore, contain the structure with half the true strength, plus an inverse image, which also has half the true strength.

Considering next the sine synthesis $i|F_N| \sin \phi_N$, we have the relation

$$i|F_N| \sin \phi_N = [\tfrac{1}{2}|F_N| \exp(i\phi_N) - \tfrac{1}{2}|F_N| \exp(-i\phi_N)]. \tag{7}$$

This is similar to the cosine synthesis considered above, but with one difference, namely that, at the inverse positions, the strengths of the peaks are negative, but equal in magnitude to $\tfrac{1}{2}f_j$.

Thus we have
cosine synthesis:

$$\tfrac{1}{2}f_j \text{ at } \pm\mathbf{r}_j (j = 1 \text{ to } N)$$

sine synthesis:

$$\left.\begin{array}{l} \tfrac{1}{2}f_j \text{ at } \mathbf{r}_j \\ -\tfrac{1}{2}f_j \text{ at } -\mathbf{r}_j \end{array}\right\} \quad (j = 1 \text{ to } N) \tag{8}$$

It is readily verified that the sum of the two gives peaks f_j at $\mathbf{r}_j$, as obviously it should, since it is equivalent to performing the complete Fourier synthesis F_N.

References

[1] G. N. Ramachandran and S. Raman. *Syntheses for the deconvolution of the Patterson function. Part I. General principles.* Acta Crystallogr. **12** (1959) 957–964.

[2] G. N. Ramachandran and R. Srinivasan. *An apparent paradox in crystal structure analysis.* Nature **190** (1961) 159–161.

[3] S. Raman. *Syntheses for the deconvolution of the Patterson function. Part II. Detailed theory for non-centrosymmetric crystals.* Acta Crystallogr. **12** (1959) 964–975.

[4] S. Raman. *Syntheses for the deconvolution of the Patterson function. Part III. Theory for centrosymmetric crystals.* Acta Crystallogr. **14** (1961) 148–150.

[5] R. Srinivasan. *The significance of the phase synthesis.* Proc. Ind. Acad. Sci. **53A** (1961) 252–261.

[6] R. Srinivasan. *Syntheses for the deconvolution of the Patterson function. Part IV. Refinement of the theory and a general comparison of various syntheses.* Acta Crystallogr. **14** (1961) 607–611.

[7] R. Srinivasan and R. Chandrasekharan. *Correlation functions connected with structure factors and their application to the observed and calculated structure factors.* Ind. J. Pure Appl. Phys. **4** (1966) 178–186.

5

Syntheses for partially known structures

Introduction

The earliest and perhaps the most commonly used method of structure analysis is the so-called heavy atom method, which depends on initially determining the positions of any heavy atom in the structure, and using the phases corresponding to these, along with the structure amplitudes of the entire crystal, in a Fourier synthesis to develop the rest of the structure. The determination of the locations of the heavy atom is usually achieved without much difficulty, say from the Patterson function. The phases calculated from the heavy atom coordinates will thus serve as a first approximation to the correct phases. We shall consider in this and the succeeding two chapters this particular type of problem, namely that of developing the full structure when a part of it is known. We shall consider this quite generally and discuss the problem when a part consisting, say, of P atoms (not necessarily heavy atoms) is known, out of a total of N atoms in the structure. Under these circumstances, the quantities that are known are $|F_N|$, the structure amplitude of the entire crystal, $F_P = |F_P| \exp(i\phi_P)$, the structure factor, both in magnitude and phase, corresponding to the P known atoms. We would like to investigate whether the usual heavy atom method, which uses $|F_N| \exp(i\phi_P)$ as coefficients in a Fourier synthesis, is the only one which could be used under these conditions to extract the rest of the structure, or whether other types of syntheses are possible, and if so, which of them is the most powerful one. Before we discuss these, we shall consider in more detail the sum function, whose Fourier representation was considered in Ch. 3.

The weighted sum function—Fourier interpretation

We saw in Ch. 3 that shifting the Patterson function by a vector, say $(\mathbf{r}_i - \mathbf{r}_1)$ is equivalent to adding a term in the Fourier synthesis of the sum function namely

$$|F_N|^2 \exp\left[2\pi i(\mathbf{r}_i - \mathbf{r}_1)\right] \qquad (1)$$

for the shifted Patterson. If, in addition to this, we multiply the contribution to the sum function for this vector shift $(\mathbf{r}_i - \mathbf{r}_1)$ by $f_i f_1$, which is the product of the scattering factors f_i and f_1 of the two atoms involved, we see that the term added to the Fourier coefficient is

$$|F_N|^2 f_1 f_i \exp\left[2\pi i(\mathbf{r}_i - \mathbf{r}_1)\right]. \qquad (2)$$

If the above process is repeated for every one of the P vectors $(\mathbf{r}_i - \mathbf{r}_1)$, $i = 1$ to P, we get, for the Fourier coefficient of the total sum function, the expression

$$f_1 \exp\left(2\pi i \mathbf{r}_1\right) |F_N|^2 \sum_{i=1}^{P} f_{Pi} \exp\left(2\pi i \mathbf{r}_{Pi}\right) = f_1 \exp\left(2\pi i \mathbf{r}_1\right)\left[|F_N|^2 F_P\right]. \qquad (3)$$

The quantity within brackets in (3) is the product of the intensity of the reflection and the structure factor for the known atoms P, and is therefore readily calculated. The first term in (3) denotes that the resultant diagram reveals the image of the structure as seen from 1 and also weighted by its scattering factor f_1. We could obviously ignore the relative shift of the image and also the multiplying factor, and it follows therefore that the synthesis $|F_N|^2 F_P$ is essentially equivalent to a "weighted" sum function and should therefore reveal the structure. We might call this the *alpha (α) synthesis*.

Interpretation of the alpha synthesis[1,3,4,7]

While we could interpret the α synthesis geometrically from the point of view of the image-seeking sum function, it is possible to work out in full detail the peaks and their positions using the convolution principle discussed in Ch. 1. It may be seen that this synthesis is a convolution of the Patterson of the entire structure, $|F_N|^2$, with F_P, which is the structure composed of the P atoms alone. The peak strengths and positions are known precisely in both of these structures and this enables us to work out the positions and strengths of the peaks in the alpha synthesis. Further we shall call this as the *alpha general synthesis* (α_{gen}) in order to distinguish it from other syntheses of the alpha class to be considered later.

We have, to start with, the relation $F_N = F_P + F_Q$ where P and Q are the known and unknown sets of atoms, respectively. Thus, following the notations mentioned in Ch. 1, P and Q will denote the number of atoms in the known and unknown part, respectively, and will also serve as labels to denote these two groups of atoms. Then,

$$|F_N|^2 = (F_P + F_Q)(\tilde{F}_P + \tilde{F}_Q) = F_P\tilde{F}_P + F_Q\tilde{F}_Q + F_P\tilde{F}_Q + \tilde{F}_P F_Q. \qquad (4)$$

Thus the α_{gen} synthesis is the structure

$$|F_N|^2 F_P = |F_P|^2 F_P + |F_Q|^2 F_P + F_P{}^2\tilde{F}_Q + |F_P|^2 F_Q. \qquad (5)$$
$$\quad\quad\quad\;\; 1 \qquad\quad\; 2 \qquad\quad\; 3 \qquad\quad 4$$

Peaks in the noncentrosymmetric case. Each term on the right-hand side of (5) will give rise to the convolution of two structures represented by the two terms involved and the positions and strengths of the peaks can be readily worked out by the method described already.

These are given in Table 1. It may be seen that the right-hand side of (5)

Table 1　Peaks in the alpha general synthesis: noncentrosymmetric case

Description	No. of peaks	Position	Strength	Designation
Known	P	$\mathbf{r}_{Pj}$	$2f_{Pj}\sum f_{Pi}^2$	1.1
Unwanted	$P(P^2 - 2P)$	$\mathbf{r}_{Pk} + \mathbf{r}_{Pi} - \mathbf{r}_{Pj}$ $(i \neq j),\,(j \neq k)$	$f_{Pi}f_{Pj}f_{Pk}$	1.2
Known	P	$\mathbf{r}_{Pj}$	$f_{Pj}\sum f_{Qi}^2$	2.1
Unwanted	$P(Q^2 - Q)$	$\mathbf{r}_{Pk} + \mathbf{r}_{Qi} - \mathbf{r}_{Qj}$ $(i \neq j)$	$f_{Qi}f_{Qj}f_{Pk}$	2.2
Unwanted	PQ	$2\mathbf{r}_{Pi} - \mathbf{r}_{Qk}$	$f_{Pi}^2 f_{Qk}$	3.1
Unwanted	$\tfrac{1}{2}(P^2 - P)Q$	$\mathbf{r}_{Pi} + \mathbf{r}_{Pj} - \mathbf{r}_{Qk}$ $(i \neq j)$	$2f_{Pi}f_{Pj}f_{Qk}$	3.2
Wanted	Q	$\mathbf{r}_{Qk}$	$f_{Qk}\sum f_{Pi}^2$	4.1
Unwatnted	$(P^2 - P)Q$	$\mathbf{r}_{Pi} - \mathbf{r}_{Pj} + \mathbf{r}_{Qk}$ $(i \neq j)$	$f_{Pi}f_{Pj}f_{Qk}$	4.2

is a sum of four terms, each one of which is the structure factor (or Fourier transform) of the convolution of two structures. The peaks arising from each of these terms can thus be written down readily, using (8) of Ch. 2. In Table 1 the peaks arising from one particular term, say 1, are denoted by 1.1, 1.2, etc., and similarly for the other terms. These peaks may be broadly classified under three categories, namely those occurring at the known atomic positions (P atoms), the wanted atomic positions (Q atoms) and those that are unwanted. The last of these do not accumulate at definite locations and constitute mainly the background in the map. Neglecting the background and using notation

$$S_P^2 = \sum_{j=1}^{P} f_{Pj}^2 \quad \text{and} \quad S_Q^2 = \sum_{j=1}^{Q} f_{Qj}^2,$$

we see that the peak strength at the known (P) atoms and the wanted (Q) atoms are, respectively, $(S_N^2 + S_P^2) f_{Pj}$ and $S_P^2 f_{Qk}$. It is convenient to normalize[†] the peak strength such that the peak strength at the known atoms P_j are made equal to f_{Pj}. Thus dividing the peak strengths at the positions of both the P and Q atoms by $(S_N^2 + S_P^2)$ we get the normalized peak strength at the wanted Q atoms to be $[S_P^2/(S_N^2 + S_P^2)] f_{Qk}$. If we further use a parameter

$$\sigma_1^2 = \frac{S_P^2}{S_N^2}, \tag{6}$$

we obtain the normalized strength at the Q atoms to be

$$\frac{\sigma_1^2}{1 + \sigma_1^2} f_{Qk}. \tag{7}$$

We may denote the normalized ratio of the peak strength at an unknown (Q) atom to that at the known (P) atoms by a quantity $\rho(f_{Qk}/f_{Pj})$. In the present example $\rho_\alpha = \sigma_1^2/(1 + \sigma_1^2)$. Since the value of ρ will be unity for a Fourier synthesis F_N (performed using the correct amplitudes $|F_N|$ and the correct phases ϕ_N), we may call ρ as the enhancement factor. This enhancement factor is, in general, less than unity, but may be greater than unity in centrosymmetric crystals for the β synthesis, as will be shown in Ch. 6.

From (7), since the value of σ_1 ranges from 0 to 1, the value of the enhancement factor ρ ranges from 0 to $\frac{1}{2}$, with increasing values of σ_1^2. Thus, even in the limit when almost all the atoms in the structure are known and are used in the α synthesis, the remaining Q atoms can occur at most with a strength of only one-half relative to the P atoms.

The quantity σ_1^2, which is defined by (6), requires special mention. If all the reflections for a given range of $(\sin \theta)/\lambda$ are considered, and the mean values of $|F_N|^2$ and $|F_P|^2$ are calculated, then the ratio $\langle |F_P|^2 \rangle / \langle |F_N|^2 \rangle$ can be shown theoretically to be equal to $S_P^2/S_N^2 = \sigma_1^2$. In fact, it is found[2,8–15] that this factor σ_1^2 plays a vital role in all the statistical formulas connecting F_N and F_P, such as the mean value of $||F_N| - |F_P||$ and so on. It also follows from the

---

† The normalized peak strength is convenient to handle in practice, since, for instance, in any practical calculation, the scaling errors in a synthesis may alter the absolute values of the peaks. The ratio of, say, the peak strength at a Q atom to that at a P atom, on the other hand, should still be comparable with the normalized peak strength at Q.

above that the mean fractional contribution to the intensity of x-ray reflections from the set of P atoms, out of the total number of N atoms, is σ_1^2. We shall find that this parameter occurs throughout this monograph in all formulas connected with the strengths of the peaks found in a Fourier synthesis obtained with partial information.

Although it can be expressed in terms of σ_1^2, it is sometimes found useful to have a symbol for $\langle |F_Q|^2 \rangle / \langle |F_N|^2 \rangle$. This is denoted by σ_2^2. Thus

$$\sigma_2^2 = \frac{S_Q^2}{S_N^2}, \tag{8a}$$

and it is obvious that

$$\sigma_1^2 + \sigma_2^2 = 1. \tag{8b}$$

Centrosymmetric case.[4] When the structure is centrosymmetric, the expressions are simpler. Thus we have the results $\tilde{F}_Q = F_Q$, and $|F_P|^2 = F_P^2$ and expression (5) reduces to

$$\alpha_{\text{gen}} = |F_N|^2 F_P = \underset{1}{|F_P|^2 F_P} + \underset{2}{|F_Q|^2 F_P} + \underset{3}{2|F_P|^2 F_Q}. \tag{9}$$

Term *3* in (9) is actually a combination of terms *3* and *4* of the noncentrosymmetric case. The peak strengths and positions of the peaks can be worked out using Table 1 of Ch. 2, and they are given in Table 2. Contributions to the

Table 2   **Peaks in the alpha general synthesis: centrosymmetric case**[†]

Description	Position	Strength	Designation
Known	$\pm \mathbf{r}_{pi}$	$2f_{pi} \sum f_{Pj}^2$	*1.1*
Unwanted	$\pm \mathbf{r}_{pi} \pm 2\mathbf{r}_{pj}$	$f_{pi} f_{pj}^2$	*1.2*
Unwanted	$\pm \mathbf{r}_{pi} \pm (\mathbf{r}_{pj} + \mathbf{r}_{pk})$	$2f_{pi} f_{pj} f_{pk}$	*1.3*
Unwanted	$\pm \mathbf{r}_{pi} \pm (\mathbf{r}_{pj} - \mathbf{r}_{pk})$	$2f_{pi} f_{pj} f_{pk}$	*1.4*
Known	$\pm \mathbf{r}_{pi}$	$f_{pi} \sum f_{Qj}^2$	*2.1*
Unwanted	$\pm \mathbf{r}_{pi} \pm 2\mathbf{r}_{qj}$	$f_{pi} f_{qj}^2$	*2.2*
Unwanted	$\pm \mathbf{r}_{pi} \pm (\mathbf{r}_{qj} + \mathbf{r}_{qk})$	$2f_{pi} f_{qj} f_{qk}$	*2.3*
Unwanted	$\pm \mathbf{r}_{pi} \pm (\mathbf{r}_{qj} - \mathbf{r}_{qk})$	$2f_{pi} f_{qj} f_{qk}$	*2.4*
Wanted	$\pm \mathbf{r}_{qi}$	$2f_{qi} \sum f_{Pj}^2$	*3.1*
Unwanted	$\pm \mathbf{r}_{qi} \pm (\mathbf{r}_{pj} + \mathbf{r}_{pk})$	$4f_{qi} f_{pj} f_{pk}$	*3.2*
Unwanted	$\pm \mathbf{r}_{qi} \pm (\mathbf{r}_{pj} - \mathbf{r}_{pk})$	$4f_{qi} f_{pj} f_{pk}$	*3.3*
Unwanted	$\pm \mathbf{r}_{qi} \pm 2\mathbf{r}_{pj}$	$2f_{qi} f_{pj}^2$	*3.4*

[†] p, q are number of atoms in asymmetric unit in the known and the unknown parts. Because of the centrosymmetry, $p = P/2$ and $q = Q/2$.

peak strengths at the known P atoms come from terms 1 and 2 and they are the same as in the noncentrosymmetric case, being equal to $S_N^2 + S_P^2$. However, the strengths of the wanted peaks at the Q atoms are doubled compared to the noncentrosymmetric case, which is due to the fact that term 3 of expression (5) for the noncentrosymmetric case, which contributed to the background, now combines with term 4 which becomes equivalent to it, and leads to the doubled strength at the Q atoms, term 3 of (9). In fact, this type of doubling will be found to be a general feature which occurs systematically in the centrosymmetric case for other syntheses also.

Effect of wrong atoms in the alpha synthesis

From the previous sections it is seen that the strengths of the peaks at the known P atoms and the wanted Q atoms involve basically the parameter σ_1^2, which measures the proportion of the mean contribution to the intensity from the known part P in relation to the mean total intensity. In actual practice, it might happen that the data regarding the input atoms, which may be indicated as the I group (I standing for "input" and I being equal to the number of atoms in the group) may contain a part, P, of atoms at correct positions ($\mathbf{r}_{Pj}$) and another part, W, in wrong positions. These latter may be denoted by strengths f_{Wj} at $\mathbf{r}_{Wj}$ of the wrong group of W atoms. The question then arises as to what the effect of these wrong atoms will be on the resultant synthesis. In particular, we are interested in the analysis of the synthesis $|F_N|^2 F_I$, where $F_I = |F_I| \exp(i\phi_I)$. This has been considered by Raman and Lipscomb.[5] We shall not give the details except to reproduce two of the tables from this work, since an outline of the method of approach for the α synthesis with wrong atoms, (as well as for the β synthesis to be considered in the next chapter) is discussed in Ch. 7. A modified form of the α synthesis, as well as an analysis of the β synthesis, have also been discussed by Sax.[6]

Table 3 shows the locations and strengths of the peaks in an α synthesis, in

Table 3   Locations and strengths of the peaks in the alpha synthesis
when the input model contains wrong atoms

Location	Normalized strength	Nature
$\mathbf{r}_{Pj}$	$(N + P - 1)$	Known, correct (P type)
$\mathbf{r}_{Wj}$	N	Input, wrong (W type)
$\mathbf{r}_{Qj}$	P	Required (Q type)
$\mathbf{r}_{Pi} - \mathbf{r}_{Nj} + \mathbf{r}_{Pk}(i \neq j, i \neq k, j \neq Pk)$	2	Background
$2\mathbf{r}_{Pi} - \mathbf{r}_{Nj}$	1	Background
$\mathbf{r}_{Ni} - \mathbf{r}_{Nj} + \mathbf{r}_{Pk}(i \neq j, i \neq Pk, j \neq Pk)$	1	Background

which the input data contain wrong atoms. For convenience, all atoms are assumed to be alike, and the normalized strengths are given as the number of contributions to each peak. It is interesting to note that the wrong atoms (W) come out with a strength distinctly less than that of the correct input atoms (P), the ratio of the two being $N/(N + P - 1)$, which, for large N is $N/(N + P)$, which is equal to $1/(1 + \sigma_1^2)$. The ratio is smaller, the larger the number of P atoms, going down to a limiting value of 0.5 when $P \to N$.

It would be interesting to find out what happens if a "difference" type of synthesis is calculated. We may consider, in particular, the following α-difference synthesis

$$(|F_N|^2 - |F_I|^2 - \sum f_{Qj}^2)|F_I| \exp(i\phi_I). \tag{10}$$

The results are shown in Table 4 in the same format as Table 3. Since there is a

Table 4   Locations and strengths of peaks in the alpha-difference synthesis, when the input model contains wrong atoms[5]

Location	Normalized strength	Nature
r_{Pj}	$-2W$	Known, correct
r_{Wj}	$-(2W + P - 1)$	Input, wrong
r_{Qj}	$+P$	Required
$r_{Pi} - r_{Qj} + r_{Pk}(i \neq k)$	$+2$	Positive background
$2r_{Pk} - r_{Qj}$	$+1$	Positive background
$r_{Qi} - r_{Qj} + r_{Ik}(i \neq j)$	$+1$	Positive background
$r_{Pi} - r_{Wj} + r_{Pk}(i \neq k)$	-2	Negative background
$2r_{Pi} - r_{Wj}$	-1	Negative background
$r_{Wi} - r_{Wj} + r_{Pk}(i \neq j)$	-1	Negative background
$r_{Wi} - r_{Wj} + r_{Wk}(i \neq j, j \neq k, i \neq k)$	-2	Negative background
$2r_{Wi} - r_{Wj}(i \neq j)$	-1	Negative background

subtraction of strength at the positions of the input atoms, this leads to negative peak strength at both P and W atoms. However, the wrong atoms now come out more strongly negative and they should be capable of being distinguished. On the other hand, the unknown required (Q) atoms come out with positive strength in the difference synthesis. (There are some errors in Ref. 5 regarding this synthesis, which have been corrected here.)

References

[1] G. N. Ramachandran and S. Raman. *Syntheses for the deconvolution of the Patterson function. Part I. General principles.* Acta Crystallogr. **12** (1959) 957–964.

[2] G. N. Ramachandran, R. Srinivasan, and V. Raghupathy Sarma. *Probability distribution connected with structure amplitudes of two related crystals. Part I. Probability distribution of the difference.* Acta Crystallogr. **16** (1963) 662–666.

[3] S. Raman. *Syntheses for the deconvolution of the Patterson function. Part II. Detailed theory for noncentrosymmetric crystals.* Acta Crystallogr. **12** (1959) 964–975.

[4] S. Raman. *Syntheses for the deconvolution of the Patterson function. Part III. Theory for centrosymmetric crystals.* Acta Crystallogr. **14** (1961) 148–150.

[5] S. Raman and W. N. Lipscomb. *Applications of Fourier transform theory to electron density extraction of Patterson functions.* Z. Kristallogr. **119** (1963) 30–41.

[6] M. Sax. *Convolution applied to the trial and error method of crystal structure analysis.* Acta Crystallogr. **16** (1963) 439–443.

[7] R. Srinivasan. *Syntheses for the deconvolution of the Patterson function. Part IV. Refinement of the theory and a general comparison of the various syntheses.* Acta Crystallogr. **14** (1961) 607–611.

[8] R. Srinivasan. *Weighting functions for use in the early stages of structure analysis when part of the structure is known.* Acta Crystallogr. **20** (1966) 143–144.

[9] R. Srinivasan and R. Chandrasekharan. *Correlation functions connected with structure factors and their application to the case of observed and calculated structure factors.* Ind. J. Pure Appl. Phys. **4** (1966) 178–186.

[10] R. Srinivasan, V. Ragupathy Sarma, and G. N. Ramachandran. *Statistical tests for isomorphism.* In "Crystallography and crystal perfection" (Ed.) G. N. Ramachandran (Academic Press, London, 1963) 85–98.

[11] R. Srinivasan, V. Ragupathy Sarma, and G. N. Ramachandran. *Probability distribution connected with structure amplitudes of two related crystals. Part II. Probability distribution of the product.* Acta Crystallogr. **16** (1963) 1151–1156.

[12] R. Srinivasan and G. N. Ramachandran. *Probability distribution connected with structure amplitudes of two related crystals. Part IV. Distribution of the normalised difference.* Acta Crystallogr. **19** (1965) 1003–1007.

[13] R. Srinivasan and G. N. Ramachandran. *Probability distribution connected with structure amplitudes of two related crystals. Part V. The effect of errors in the atomic coordinates on the distribution of observed and calculated structure factors.* Acta Crystallogr. **19** (1965) 1008–1014.

[14] R. Srinivasan and G. N. Ramachandran. *Probability distribution connected with structure amplitudes of two related crystals. Part VI. On the significance of the parameter σ_A.* Acta Crystallogr. **20** (1966) 570–571.

[15] R. Srinivasan, E. Subramanian, and G. N. Ramachandran. *Probability distribution connected with structure amplitudes of two related crystals. Part III. Probability distribution of quotient.* Acta Crystallogr. **17** (1964) 1010–1014.

6

Syntheses for partially known structures: other types of syntheses

General form of the syntheses

The interpretation of the alpha synthesis based on the convolution principle, discussed in the last chapter, enables us to extend the argument in order to search for other possible syntheses which could extract the full structure when a part of it is known. In Ch. 2 we have seen that, in general, syntheses with coefficients $|F_N|^n$ have properties similar to those of the Patterson function. We have also seen in Ch. 4 that the three syntheses

$$F_P = |F_P| \exp(i\phi_P), \quad \exp(i\phi_P) \quad \text{and} \quad (1/\tilde{F}_P) = (1/|F_P|) \exp(i\phi_P)$$

are also similar, all of them having their principal peaks in the same positions, namely those corresponding to the atoms in the true structure F_P. A reference to Table 1 of Ch. 5 shows clearly that the peaks at the wanted Q atoms develop essentially as a consequence of the convolution of the Patterson part $|F_N|^2$ with the true structure of the P atoms represented by F_P. Consequently, if, in this synthesis, $|F_N|^2$ is replaced by $|F_N|^n$, and F_P is replaced by one of F_P, $\exp(i\phi_P)$ or $(1/\tilde{F}_P)$, the resultant synthesis would be expected to have properties similar to the α synthesis and reveal concentrated peaks at positions corresponding to the unknown Q atoms. We shall however restrict ourselves to the cases when the exponent n in $|F_N|^n$ is equal to 2 or 1. The resultant two series of syntheses may be called[†] the α, β, γ and α', β', γ' syntheses,[1-4] defined by the Fourier coefficients in (1)–(6).

[†] Strictly, these are general synthesis denoted by α_{gen}, β_{gen}, etc., and other modified forms of these will be discussed later. The subscript "gen" is optionally omitted when no confusion is likely to arise.

α synthesis:

$$|F_N|^2 F_P = |F_N|^2 |F_P| \exp(i\phi_P) \tag{1}$$

β synthesis:

$$\frac{|F_N|^2}{\tilde{F}_P} = \frac{|F_N|^2}{|F_P|} \exp(i\phi_P) \tag{2}$$

γ synthesis:

$$|F_N|^2 \exp(i\phi_P) \tag{3}$$

α' synthesis:

$$|F_N| F_P = |F_N| |F_P| \exp(i\phi_P) \tag{4}$$

β' synthesis:

$$\frac{|F_N|}{\tilde{F}_P} = \frac{|F_N|}{|F_P|} \exp(i\phi_P) \tag{5}$$

γ' synthesis:

$$|F_N| \exp(i\phi_P) \tag{6}$$

Considering the various syntheses from which these are obtained, only the peaks in the Patterson $|F_N|^2$ and the Fourier F_P are exactly and fully describable. The other structures, such as those arising from $|F_P|$, $\exp(i\phi_P)$ and $(1/\tilde{F}_P)$, can be obtained only approximately. Consequently, of the six syntheses (1)–(6) mentioned above, only the first one, namely the α synthesis, can be described exactly, while the strengths of only the principal peaks in the others can be theoretically obtained in a simple manner. Of these others, two are of special interest, namely the β and γ' syntheses, and these will be discussed in good detail. These two have the property that, as P tends to N, the coefficients tend to a value F_N, the true structure factors of the full crystal. The synthesis γ' is readily seen to be nothing but the conventional heavy-atom synthesis. It may be mentioned that a synthesis of the beta type was earlier suggested by Rogers,[5] but no detailed analysis of how it could lead to the structure was made. As regards the heavy-atom synthesis, Luzzati[1] was the first to show, using statistical analysis, that the synthesis should reveal the rest of the structure. Our present analysis is based on the method of approach developed by the Madras group,[2–4,6] which shows that these syntheses do lead to the development of the rest of the structure. In particular, it is found that the β synthesis could lead to higher peak strengths at the unknown atomic positions than the γ' synthesis and that it could also suppress wrong atoms fed into the input data more effectively.

The beta syntheses

We expand the coefficient $|F_N|^2/\tilde{F}_P$ of the beta synthesis, using $F_N = F_P + F_Q$, and obtain it in the form

$$\beta_{\text{gen}} = \frac{|F_N|^2}{\tilde{F}_P} = \frac{F_P\tilde{F}_P + F_Q\tilde{F}_Q + F_P\tilde{F}_Q + \tilde{F}_PF_Q}{\tilde{F}_P}, \tag{7a}$$

$$= F_P + \frac{|F_Q|^2}{\tilde{F}_P} + \tilde{F}_Q\exp(2i\phi_P) + F_Q. \tag{7b}$$

$$\quad\quad 5 \quad\quad\quad 6 \quad\quad\quad 7 \quad\quad\quad 8$$

The symbol β_{gen} is used to distinguish it from other β type syntheses to be described below. We can work out the strengths and positions of the peaks arising out of each one of the four terms in (7b) in a manner similar to that adopted for the α synthesis.

It may be noticed that two out of the four terms in (7), F_P and F_Q, terms 5 and 8, respectively, lead exactly to the peaks at the P and Q atomic positions, with the correct strengths. The other two involve $\exp(2i\phi_P)$ and $(1/\tilde{F}_P)$, for whose structures we can only obtain an approximation. On taking the first-order approximation for these two, as discussed in Ch. 4, we arrive at the results shown in Table 1. The peak positions are found to be the same as in

Table 1   List of peaks in the beta general synthesis for a noncentrosymmetric structure

No. of peaks	Position	Strength	Designation	Corresp. peak in α_{gen} synthesis
P	$\mathbf{r}_{Pj}$	f_{Pj}	5.1	1.1
P	$\mathbf{r}_{Pj}$	$(S_Q^2/S_P^2)f_{Pj}$	6.1	2.1
$P(Q^2-Q)$	$\mathbf{r}_{Pk}+\mathbf{r}_{Qi}-\mathbf{r}_{Qj}$ $(i\neq j)$	$(1/S_P^2)f_{Pk}f_{Qi}f_{Qj}$	6.2	2.2
$P(P^2-P)$	$\mathbf{r}_{Pk}+\mathbf{r}_{Pi}-\mathbf{r}_{Pj}$ $(i\neq j)$	$-(S_Q^2/S_P^2)f_{Pi}f_{Pj}f_{Pk}$	6.3	1.2
PQ	$2\mathbf{r}_{Pi}-\mathbf{r}_{Qk}$	$(1/S_P^2)f_{Pi}^2f_{Qk}$	7.1	3.1
$(P^2-P)Q$	$\mathbf{r}_{Pi}+\mathbf{r}_{Pj}-\mathbf{r}_{Qk}$	$(2/S_P^2)f_{Pi}f_{Pj}f_{Pk}$	7.2	3.2
Q	$\mathbf{r}_{Qk}$	f_{Qk}	8.1	4.1

α_{gen} synthesis excepting that there are no peaks corresponding to 1.2 and 4.2 of that synthesis. The correspondence of the other terms giving rise to various peaks in the two syntheses may be seen from the last two columns of Table 1. The strengths at the known (P) and the wanted (Q) positions are seen to be respectively equal to $(S_N^2/S_P^2)f_{Pj}$ and f_{Qk}. The normalized peak strength at the wanted Q atoms is thus equal to $\sigma_1^2 f_{Qk}$ which is greater than the corresponding value for the α_{gen} synthesis, namely, $[\sigma_1^2/(1 + \sigma_1^2)]f_{Qk}$. In particular, as P tends to N, the normalized strength at the Q atoms tends to 1, i.e., the required unknown atoms come out with their full strength.

The absence of terms corresponding to *1.2* and *4.2* of the α_{gen} synthesis shows that the background in the beta synthesis is less than that in the α_{gen} synthesis. Another important feature to be noticed is that the dispersed peaks *6.3* are of negative strength. These would also partly cancel the background due to the positive peaks and thus improve the general background. These features indicate that the beta synthesis is better than the alpha synthesis in revealing the unknown part of the structure. In fact, anticipating our discussion of the heavy atom synthesis in the next section, the beta synthesis turns out to be the most powerful of the three syntheses α, β, and γ'.

There is however one important precaution that is necessary in the use of the beta synthesis. It stems from the fact that unlike the α and the γ' syntheses, in which the Fourier coefficient is the product of finite quantities, the coefficient used in the β synthesis is the ratio of $|F_N|^2$ to $|F_P|$, multiplied by a phase factor. Thus it contains a quantity $|F_P|$ in the denominator. This would mean that whenever $|F_P|$ is small and tends to zero, the corresponding coefficient is large and tends to infinity. Consequently, the Fourier synthesis of the β structure will be dominated by these terms, and this will lead to a widely fluctuating background. It is however possible to avoid this difficulty by omitting from the synthesis those terms for which $|F_P|$ is less than a given fraction of the mean value of $|F_P|$ for the particular range of $(\sin \theta)/\lambda$. The details of such procedures of weighting the terms will be discussed in Ch. 8. However, it is obvious that the omission of such terms with small $|F_P|$ from the Fourier synthesis would not produce a serious error, since the phase information provided by the input data F_P will not be of much significance in such cases. Obviously, if the cut off limit for $|F_P|$ is made too high, then the advantage of the β synthesis is lost. This again will be discussed in Ch. 8 in connection with the experimental verification of these ideas.

To summarize the main results regarding the β synthesis, the peak strengths expected in this synthesis are

$$\frac{S_N^2}{S_P^2} f_{Pj} \quad \text{at the known atoms} \quad \mathbf{r}_{Pj}. \tag{8a}$$

and

$$f_{Qk} \text{ at the unknown atoms } \mathbf{r}_{Qk}. \tag{8b}$$

Thus the enhancement factor is, in general, less than unity. It is in fact given by

$$\rho_{P\beta} = \frac{S_P^2}{S_N^2} = \sigma_1^2. \tag{9}$$

which tends to 1, as $P \rightarrow N$, which is to be expected.

The gamma-prime synthesis

We shall now consider the γ' synthesis, which is the conventional heavy-atom synthesis. It is interesting that this has been used right from the early days of structure analysis. However, the validity of the approach was based more on simple physical reasoning, rather than on any formal theoretical proof that it should yield the rest of the structure. Thus it is clear that as $P \to N$, this combination $|F_N| \exp(i\phi_P)$ tends to have the correct phase, namely ϕ_N. However, a proof that the γ' synthesis has peaks of appreciable strength at the unknown atomic positions comes out in a straightforward way, in a manner very similar to that adopted for the β synthesis, from our present approach. However, Luzzati[1] was able to show, using an entirely different statistical argument, that a heavy-atom synthesis yields the rest of the structure. Our present approach is however much simpler and gives more quantitative values for the peak strengths.

Taking the results to the first order of approximation for both the modulus synthesis, $|F_N|$, and the phase synthesis, $\exp(i\phi_P)$, the peaks in the γ' synthesis, $|F_N| \exp(i\phi_P)$ can readily be worked out and are given in Table 2. The peak

Table 2 List of peaks in the gamma-prime synthesis for a noncentrosymmetric structure

No. of peaks	Position	Strength	Designation	Corresponding peak in α_{gen} synthesis
P	$\mathbf{r}_{Pj}$	$(S_N/S_P)f_{Pj}$	9.1	1.1
P	$\mathbf{r}_{Pj}$	$-(S_N/2S_P)f_{Pj}$	9.2	—
P	$\mathbf{r}_{Pj}$	$(S_P/2S_N)f_{Pj}$	10.1	2.1
$P(P^2 - P)$	$\mathbf{r}_{Pk} + \mathbf{r}_{Pl} - \mathbf{r}_{Pj}$ $(i \neq j, k \neq j)$	$(1/2S_N S_P)f_{Pl}f_{Pj}f_{Pk}$	10.2	1.2
$P(Q^2 - Q)$	$\mathbf{r}_{Pk} + \mathbf{r}_{Ql} - \mathbf{r}_{Qj}$ $(i \neq j)$	$(1/2S_N S_P)f_{Ql}f_{Qj}f_{Pk}$	11.1	2.2
PQ	$2\mathbf{r}_{Pl} - \mathbf{r}_{Qk}$	$(1/2S_N S_P)f_{Pl}^2 f_{Qj}$	12.1	3.1
$\frac{1}{2}(P^2 - P)Q$	$\mathbf{r}_{Pl} + \mathbf{r}_{Pj} - \mathbf{r}_{Qk}$ $(i \neq j)$	$(1/S_N S_P)f_{Pl}f_{Pj}f_{Qk}$	12.2	3.2
Q	$\mathbf{r}_{Qk}$	$(S_P/2S_N)f_{Qk}$	13.1	4.1
$(P^2 - P)Q$	$\mathbf{r}_{Pl} - \mathbf{r}_{Pj} + \mathbf{r}_{Qk}$ $(i \neq j)$	$(1/2S_N S_P)f_{Pl}f_{Pj}f_{Qk}$	13.2	4.2

strengths at the known and unknown atomic positions are

$$\frac{S_N^2 + S_P^2}{2S_P S_N} f_{Pj} \quad \text{at} \quad \mathbf{r}_{Pj}, \tag{10a}$$

and

$$\frac{S_P}{2S_N} f_{Qk} \quad \text{at} \quad \mathbf{r}_{Qk}. \tag{10b}$$

The normalized peak strength at the required positions (Q atoms) is therefore seen to be $[\sigma_1^2/(1 + \sigma_1^2)]f_{Qj}$, so that the enhancement factor ρ_Q is

$$\rho_{Q\gamma'} = \frac{\sigma_1^2}{1 + \sigma_1^2}. \tag{11a}$$

For comparison, we may obtain the factor ρ_Q for the α synthesis from (7) of Ch. 5 to be

$$\rho_{Q\alpha} = \frac{\sigma_1^2}{1 + \sigma_1^2}. \tag{11b}$$

Thus, of the three syntheses α, β, and γ', the beta synthesis yields the required Q atoms in the structure with larger peak strength (relative to the known atoms) than the other two.

The gamma general, alpha prime, and beta prime syntheses

We shall not consider these syntheses in detail here. It is clear that, in principle, the peaks can be worked out in all these syntheses. Of these, in the γ_{gen} synthesis, namely $|F_N|^2 \exp(i\phi_P)$, one of the terms in its coefficient, $|F_N|^2$, has an exact interpretation, while $\exp(i\phi_P)$ alone involves an approximation. Both α' and β' on the other hand involve terms none of which are exactly interpretable. The peaks in γ_{gen} synthesis are given in Table 3, and

Table 3   List of peaks in γ_{gen} synthesis for a noncentrosymmetric structure

Position	Strength	Designation	Corresponding peak in	
			α_{gen}	β_{gen}
$\mathbf{r}_{PJ}$	$\frac{3}{2}S_P f_{PJ}$	*14.1*	*1.1*	*5.1*
$\mathbf{r}_{Pi} + \mathbf{r}_{PJ} - \mathbf{r}_{Pk}$ $(j \neq k)$	$(1/2S_P)f_{Pi}f_{PJ}f_{Pk}$	*14.2*	*1.2*	—
$\mathbf{r}_{PJ}$	$(S_Q^2/S_P)f_{PJ}$	*15.1*	*2.1*	*6.1*
$\mathbf{r}_{Pk} + \mathbf{r}_{Qi} - \mathbf{r}_{QJ}$ $(i \neq j)$	$(1/S_P)f_{Pk}f_{Qi}f_{QJ}$	*15.2*	*2.2*	*6.2*
$\mathbf{r}_{PJ} + \mathbf{r}_{Pk} - \mathbf{r}_{Pl}$ $(k \neq l)$	$-(S_Q^2/2S_P^3)f_{PJ}f_{Pk}f_{Pl}$	*15.3*	—	*6.3*
$2\mathbf{r}_{PJ} - \mathbf{r}_{Qk}$	$(1/S_P)f_{PJ}^2 f_{Qk}$	*16.1*	*3.1*	*7.1*
$\mathbf{r}_{Pi} + \mathbf{r}_{PJ} - \mathbf{r}_{Qk}$ $(i \neq j)$	$(2/S_P)f_{Pi}f_{PJ}f_{Qk}$	*16.2*	*3.2*	*7.2*
$\mathbf{r}_{QJ}$	$S_P f_{QJ}$	*17.1*	*4.1*	*8.1*
$\mathbf{r}_{Pi} - \mathbf{r}_{PJ} + \mathbf{r}_{Qk}$ $(i \neq j)$	$(1/2S_P)f_{Pi}f_{PJ}f_{Qk}$	*17.2*	*4.2*	—

follows the report of Srinivasan.[6] The terms involved in this synthesis are given in the right-hand side of (12)

$$\gamma_{\text{gen}} = (F_P + F_Q)(\tilde{F}_P + \tilde{F}_Q) \exp(i\phi_P),$$

$$= |F_P|^2 \exp(i\phi_P) + |F_Q|^2 \exp(i\phi_P) + |F_P| \exp(2i\phi_P)\tilde{F}_Q + |F_P| F_Q. \quad (12)$$
$$\underset{14}{} \qquad\qquad \underset{15}{} \qquad\qquad \underset{16}{} \qquad\qquad \underset{17}{}$$

The modified syntheses

A reference to Tables 1 and 3 and also to Table 1 of Ch. 5 shows clearly that apart from the peaks at the P and Q atoms, the syntheses have a good deal of unwanted background, particularly in the α type of syntheses. This feature could be more readily understood in relation to the α synthesis. Since it is a weighted sum function, it contains, besides the peaks corresponding to the image of the structure, other peaks also which are accumulated. The wanted peaks and the peaks at the P atoms only are enhanced relative to the background.

It is possible to make the syntheses more efficient by getting rid of some of the background peaks. Thus in the α synthesis, the peaks *1.2* only contribute to the background and can easily be removed since they arise only from the interactions between the P atoms. Thus, to start with, we could subtract the Patterson of the P atoms from $|F_N|^2$ and take the coefficients as $(|F_N|^2 - |F_P|^2)$. This would also result in the removal of the peaks at the positions of the P atoms, which, however, is not serious since they are known atoms. The strengths of the background peaks arising from 2 of (5) of Ch. 5 can be removed if $|F_Q|^2$ is also subtracted. However, since $|F_Q|^2$ is an unknown quantity the best that could be done is to subtract its mean value, namely $\langle|F_Q|^2\rangle = \sum f_Q^2$ and this may be expected to reduce the background arising from these terms. Thus we could consider a modified synthesis using the coefficients

$$\alpha_{\text{mod}} = (|F_N|^2 - |F_P|^2 - \sum f_{Qj}^2)F_P. \quad (13)$$

This synthesis will have the known atoms P completely suppressed and also have a certain amount of the background removed. We shall refer to the above synthesis as the alpha modified synthesis, α_{mod}. The modified synthesis of the beta and gamma classes follows in a similar way. These are

$$\beta_{\text{mod}} = \frac{|F_N|^2 - |F_P|^2 - \sum f_{Qj}^2}{\tilde{F}_P}, \quad (14)$$

$$\gamma_{\text{mod}} = (|F_N|^2 - |F_P|^2 - \sum f_{Qj}^2) \exp(i\phi_P). \quad (15)$$

Following the above arguments, we could also define the modified syntheses of the primed series in a somewhat similar manner, although the argument cannot be an exact one. Thus, considering the γ' synthesis, the modified synthesis is defined to be

$$\gamma'_{\text{mod}} = (|F_N| - |F_P|) \exp(i\phi_P). \tag{16}$$

This may be readily recognized to be the difference Fourier synthesis. This corresponds to an exact removal of the P atoms from the γ' synthesis. The other two modified syntheses could therefore be defined analogously as

$$\alpha'_{\text{mod}} = (|F_N| - |F_P|)F_P, \tag{17}$$

$$\beta'_{\text{mod}} = \frac{|F_N| - |F_P|}{|F_P|} \exp(i\phi_P). \tag{18}$$

An important precaution is necessary in using these modified syntheses, for the intensities $|F_N|^2$ derived from the set of observed data should be on the absolute scale. Since the quantity $|F_P|$, or $|F_P|^2$ as the case may be, is subtracted, any errors in scale factor in the observed intensities will affect the Fourier map. Such an absolute scale factor is not needed for any of the unmodified syntheses. So also, while subtracting $|F_P|^2$ and $\sum f_{Qj}^2$, care should be taken to correct these for temperature factors, since, normally, $|F_N|^2$ is taken to be $|F_{\text{obs}}|^2$.

The effect of centrosymmetry on beta and gamma prime syntheses

The details of the peak strengths and their positions for a centrosymmetric crystal in the case of the α synthesis were given in Ch. 4. It was then pointed out that, because of the presence of a center of symmetry, some of the peaks arising out of the various interactions merge to give an enhancement of peak strength, with a consequent reduction in the number of peaks in the background. In particular, the Q atoms come out with double the strength, as compared with the noncentrosymmetric case. This feature is, in fact, found to be present in all the other syntheses also.

We may consider in particular the β and γ' syntheses. In the case of the β synthesis, the term $\tilde{F}_Q \exp(2i\phi_P)$ in (7b) becomes the same as F_Q since $\tilde{F}_Q = F_Q$ and $\phi_P = 0$ or π in the centrosymmetric case. Thus the beta general synthesis for a centrosymmetric crystal reduces to

$$\beta_{\text{gen}} = F_P + \frac{|F_Q|^2}{\tilde{F}_P} + 2F_Q. \tag{19}$$

The Q atoms therefore come out with twice the strength of the corresponding noncentrosymmetric case, namely $2f_{Qk}$, while the P atoms continue to have strengths $(S_N^2/S_P^2)f_{Pj}$. Hence, the factor ρ_Q has a value

$$\rho_{Q\beta} = 2\sigma_1^2 \tag{20}$$

for centrosymmetric case. It is to be noted that this can be greater than unity, and that it takes a value 2 when $P \to N$, i.e., when $\sigma_1^2 \to 1$.

We shall not give the detailed theory for the γ' synthesis, but only the expression for the peak strengths of the known and unknown atoms. The theory is only approximate, and depends on the order of approximation to which the calculations are carried out. To the first order, the strengths at the known and unknown atoms come out to be the same when normalized—thus their absolute strengths are $(S_P/S_N)f_{Pj}$ and $(S_P/S_N)f_{Qk}$, respectively. The enhancement factor is thus approximately

$$\rho_{Q\gamma'} = 1. \tag{21}$$

The approximations are particularly bad for small σ_1^2, but it is seen from a comparison of (20) and (21) that when $\sigma_1^2 > \frac{1}{2}$, the beta synthesis reveals the unknown atoms with larger strength than the gamma-prime synthesis. In the limit, when $\sigma_1^2 \to 1$, i.e., when most of the atoms are known, the remaining atoms come out with the same strength as the known atoms in the γ' synthesis, while they are revealed with twice this strength in the β synthesis. A verification of these predictions is described in Ch. 8.

References

[1] V. Luzzati. *Resolution d'une structure cristalline lorsque les position d'une partie des atomes sont connues: Traitment statistique.* Acta Crystallogr. **6** (1953) 142–152.

[2] G. N. Ramachandran and S. Raman. *Syntheses for the deconvolution of the Patterson function. Part. I. General principles.* Acta Crystallogr. **12** (1959) 957–964.

[3] S. Raman. *Syntheses for the deconvolution of the Patterson function. Part II. Detailed theory for noncentrosymmetric crystals.* Acta Crystallogr. **12** (1959) 964–975.

[4] S. Raman. *Syntheses for the deconvolution of the Patterson function. Part III. Theory for centrosymmetric crystals.* Acta Crystallogr. **14** (1961) 148–150.

[5] D. Rogers. *New methods of direct structure determination using modified Patterson maps.* Research (London) **4** (1951) 295–296.

[6] R. Srinivasan. *Syntheses for the deconvolution of the Patterson function. Part IV. Refinement of the theory and a general comparison of the various syntheses.* Acta Crystallogr. **14** (1961) 607–611.

7

Syntheses for partially known structures: second-order approximation and effect of wrong atoms

Calculation to the second order of approximation

As mentioned in Ch. 5, the properties of the various syntheses, particularly regarding the strengths of the peaks, could be worked out only as successive approximations, except for the α synthesis whose Fourier coefficient contains terms which can be interpreted exactly. In the earlier chapters, the calculations involving only the first approximation for the various standard syntheses have been described. It is however possible to work out the strengths of the peaks to higher orders of approximation.[4] This does not affect the positions of the peaks, but only their strengths. We shall briefly outline the method here and discuss in detail the results with regard to the two syntheses which are of primary interest to us, namely the β and γ' syntheses.

Refinement of the standard syntheses

The interpretation of the standard syntheses in the earlier chapters was essentially based on the initial information regarding the peaks in the modulus synthesis. This was achieved by first expanding $|F|$ in a form resembling the Taylor series expansion. We must therefore refine first the modulus synthesis itself which is used in deducing the results for the other syntheses. The procedure adopted is as follows. Since the positions of the peaks are unaffected but only their strengths, we assume, to start with, that the strengths of peaks in the modulus syntheses are αS_N at the origin and $\beta f_i f_j / S_N$ at $\mathbf{r}_i - \mathbf{r}_j$, where α and β are unknowns to be determined.[†]

We now make use of the equation

$$|F_N| \, |F_N| = |F_N|^2, \tag{1}$$

and equate the first- and second-order peaks on both sides of the equation. The strengths of the peaks for the Patterson $|F_N|^2$ are known exactly, while those of $|F_N|$ on the left-hand side involve the unknown parameters α and β. Thus the contribution to the origin term from the left-hand side arises from two types of interactions between the $|F|$ structures: (a) those between the origin peaks αS_N, and (b) those between the nonorigin peaks $\beta f_i f_j / S_N$. The total contribution to the origin term on the left-hand side can be shown to be[‡] $(\alpha^2 + \beta^2) S_N^2$ and this is equated to the corresponding term on the right-hand side, giving the equation

$$\alpha^2 + \beta^2 = 1. \tag{2}$$

A similar argument for the peak strengths given by both sides at the position $(\mathbf{r}_i - \mathbf{r}_j)$ leads to the equation

$$2\beta(\alpha + \beta) = 1. \tag{3}$$

The solution of these two equations leads to the values of $\alpha = 0.924$ and $\beta = 0.383$. Taking the first approximation to α and β is equivalent to solving the two equations $\alpha^2 = 1$, $2\alpha\beta = 1$, which are obtained by taking only terms up to the first order in (1). These lead to the values $\alpha = 1.0$ and $\beta = 0.5$ used in the earlier chapters.

The refined values of α and β may be taken as the starting point for refining the other standard syntheses. It may not be worthwhile to go to still higher orders of approximation since the Taylor series is only asymptotically convergent. The second-order approximation of the peak strengths in other standard syntheses such as $\exp(i\phi)$, $\exp(2i\phi)$, etc., may be obtained by first setting up simple equations involving these and the modulus synthesis. The strengths in each of them is expressed in terms of parameters unknown to start with and these are then solved for in a manner similar to that adopted above for the modulus synthesis. However, by its nature, the problem is not capable of having a unique solution and only a "best fit" for the unknown parameter can be arrived at.[4]

[†] The quantities α, β, γ, etc., appearing in various equations in this chapter are unknown constants and should not be confused with those used elsewhere in this monograph to denote different types of syntheses.

[‡] There is a small approximation involved in writing the second term as βS_N^2, in that $\sum_i \sum_j f_i^2 f_j^2 (i \neq j)$ is put equal to $(\sum_j f_j^2)^2$. The error is not appreciable for large N.

Thus we have the following equations to start with, connecting the different standard syntheses:

$$|F| \exp(i\phi) = F, \tag{4}$$

$$\exp(i\phi) \exp(-i\phi) = 1, \tag{5}$$

$$|F|^2 \exp(i\phi) = |F| F, \tag{6}$$

$$|F|^2(\exp i\phi) = F^2 \exp(-i\phi), \tag{7}$$

$$|F|^2 \exp(2i\phi) = F^2, \tag{8}$$

$$\exp(i\phi) \exp(i\phi) = \exp(2i\phi), \tag{9}$$

$$|F| \exp(2i\phi) = F \exp(i\phi). \tag{10}$$

Since the types of peaks in each one of these are known, their peak strengths alone are taken to be unknown, as given in Table 1, where α, β, x, y, etc., are parameters to be determined. Each one of the equations above provides us with a set of equations relating these unknown constants, which are valid only to the second degree of approximation. The best fit obtained by trial and error is given in Table 1.

Table 1 Refinement of peak strengths in standard syntheses[4]

Synthesis	Peak strength	Position	Value of the unknowns	Value in the first approximation
$|F|$	αS_N	Origin	$\alpha = 0.92$	1.0
	$\beta f_i f_j / S_N$	$\mathbf{r}_i - \mathbf{r}_j$	$\beta = 0.38$	0.5
$\exp(i\phi)$	$x f_i / S_N$	$\mathbf{r}_i$	$x = 0.90$	1.0
	$y f_i f_j f_k / S_N^3$	$\mathbf{r}_i + \mathbf{r}_j - \mathbf{r}_k$	$y = -0.45$	-0.5
$\exp(2i\phi)$	$X f_i f_j / S_N^2$	$\mathbf{r}_i + \mathbf{r}_j$	$X = 0.85$	1.0
	$Y f_i f_j f_k f_l / S_N^4$	$\mathbf{r}_i + \mathbf{r}_j - \mathbf{r}_k + \mathbf{r}_l$	$Y = -2.50$	—[†]
$1/|F|$	γ / S_N	Origin	$\gamma = 1.2$	1.0
	$\delta f_i f_j / S_N^3$	$\mathbf{r}_i - \mathbf{r}_j$	$\delta = -0.27$	-0.5
$1/F$	$p f_i / S_N^2$	$-\mathbf{r}_i$	$p = 0.95$	1.0
	$q f_i f_j f_k / S_N^4$	$-\mathbf{r}_i - \mathbf{r}_j + \mathbf{r}_k$	$q = -0.66$	-1.0

[†] Y does not occur in the first-order calculation.

Table 2   Ratio of peak strengths at the wanted (Q) and the
known (P) atomic positions[4]

Synthesis	Enhancement factor ρ	
	First approximation	Second approximation
α	$\sigma_1^2/(1 + \sigma_1^2)$	$\sigma_1{}^2/(1 + \sigma_1^2)$
β	σ_1^2	$\sigma_1{}^2/(0.95 + 0.05\sigma_1^2)$
γ'	$2\sigma_1^2/(1 + \sigma_1^2)$	$\sigma_1{}^2/(2.8 + 0.43\sigma_1^2)$[†]

[†] This requires revision.

Peak strength in the beta, gamma prime, and other syntheses

Using the refined values of the peak strengths in the standard syntheses, it is possible to work out more accurate values of the strengths of the peaks in the various classes of syntheses such as α, β, γ', etc. However, we shall not work these out for all the peaks, but only for those at the known (P) and wanted (Q) positions, which alone are of particular interest to us. The ratio of the peak strengths at the Q and the P atoms, namely $\rho(f_{Qj}/f_{Pk})$ for the three syntheses,[†] α, β, and γ', are given in Table 2. For convenience of reference, the values of the enhancement factor[‡] ρ are also given to the first-order approximation. It may be noticed that the refined values are only slightly different from the original values, except for the γ' synthesis, for which the second approximation value requires reexamination.

Reference to Table 2 shows that the beta synthesis is distinctly superior to the others, and that the value of ρ is greatest for this synthesis. In the limit when $P \to N$, the value of ρ tends to unity, while it is less in the case of the other syntheses. Caution is necessary, however, when applying the criterion of ρ. The value of ρ given here refers only to the case when the background strength is zero. In an actual case, this may not be so, since the number of peaks contributing to the background may be quite large, even though individually the strength of each peak may be small. The effect is particularly serious in projections, since a large number of them can add up to enhance the background strength. The value of ρ in such cases should obviously be taken to be the peak height over and above the background.

[†] We shall be concerned only with these three syntheses for reasons previously mentioned. Details regarding the other syntheses are not given here, but they are available in the paper cited above.

[‡] We shall omit the subscript Q in ρ_Q hereafter, where the meaning is obvious.

Effect of wrong atoms[1-3]

In Ch. 5 we considered the effect of including wrong atoms in the input data used for an α synthesis. In this section we shall give the corresponding theoretical results for the β synthesis, both for noncentrosymmetric and centrosymmetric crystals.[1] However, we shall not give the full derivation in all the cases, nor give the peak strengths at other than the locations of particular interest, namely the positions of the known correct (P) atoms, the required (Q) atoms, and the wrong input (W) atoms. Following the notation used in Ch. 5, we shall use the symbol I (input) to indicate all the atoms used in the input data. Clearly, $I = P + W$.

Using the same notation as in Ch. 5 and 6, we shall denote the ratio $S_P^2/S_N^2(=\sum_1^P f_{Pi}^2/\sum_1^N f_{Nj}^2)$ by σ_1^2. In the same way, we shall use the symbols χ_1^2 and χ_2^2 to represent the quantities given in (11), where $S_I^2 = \sum_1^I f_{Ik}^2$:

$$\chi_1^2 = \frac{S_P^2}{S_I^2},\tag{11a}$$

$$\chi_2^2 = \frac{S_W^2}{S_I^2}.\tag{11b}$$

We shall also denote the enhancement factors for the Q atoms and W atoms by ρ_Q and ρ_W. Using this notation, the peak strengths and enhancement factors for the syntheses are given[1] in Table 3.

Table 3 Normalized peak strengths and enhancement factors in the alpha and beta syntheses with wrong atoms

Synthesis†		Peak strengths at			Enhancement factor	
		P atoms	Q atoms	W atoms	ρ_Q	ρ_W
α	A	$(S_N^2 + S_P^2)f_{PJ}$	$S_P^2 f_{QJ}$	$S_N^2 f_{WJ}$	$\dfrac{\sigma_1^2}{1+\sigma_1^2}$	$\dfrac{1}{1+\sigma_1^2}$
	C	$(S_N^2 + 2S_P^2)f_{PJ}$	$2S_P^2 f_{QJ}$	$S_N^2 f_{WJ}$	$\dfrac{2\sigma_1^2}{1+2\sigma_1^2}$	$\dfrac{1}{1+2\sigma_1^2}$
β	A	$\dfrac{\chi_1^2}{\sigma_1^2}f_{PJ}$	$\chi_1^2 f_{QJ}$	$\dfrac{(1-\sigma_1^2)}{\sigma_1^2}\chi_1^2 f_{WJ}$	σ_1^2	$1-\sigma_1^2$
	C	$\dfrac{\chi_1^2}{\sigma_1^2}f_{PJ}$	$2\chi_1^2 f_{QJ}$	$\dfrac{(1-2\sigma_1^2)}{\sigma_1^2}\chi_1^2 f_{WJ}$	$2\sigma_1^2$	$1-2\sigma_1^2$

† $A =$ acentric (noncentrosymmetric); $C =$ centrosymmetric.

Derivation for the centrosymmetric alpha synthesis

In order to illustrate the method of approach, we shall consider in some detail this particular case. For a general noncentrosymmetric case, the α synthesis with wrong atoms is $|F_N|^2 F_I$, where $F_N = F_P + F_Q$ and $F_I = F_P + F_W$. Substituting these, the synthesis is seen to consist of the terms in (12) below:

$$|F_P|^2 F_P + |F_Q|^2 F_P + |F_P|^2 F_Q + F_P^2 \tilde{F}_Q$$

$$+ |F_P|^2 F_W + |F_Q|^2 F_W + F_P \tilde{F}_Q F_W + \tilde{F}_P F_Q F_W. \quad (12)$$

For a centrosymmetric crystal, all the F's are real quantities, so that $\tilde{F} = F$ and (12) reduces to

$$F_P^3 + F_Q^2 F_P + 2F_P^2 F_Q + F_P^2 F_W + F_Q^2 F_W + 2F_P F_Q F_W. \quad (13)$$

In this, the only terms which lead to peaks at the P atoms are F_P^3, and the interaction of the origin peak of F_Q^2 with F_P. This is seen to give rise to a peak of strength

$$(3S_P^2 + S_Q^2) f_{Pi} = (S_N^2 + 2S_P^2) f_{Pi}.$$

Similarly, the only interaction that leads to a peak at a Q atom is that between the origin peak of F_P^2 and F_Q, in the term $2F_P^2 F_Q$, which yields a peak strength $2S_P^2 f_{Qj}$. Both the terms $F_P^2 F_W$ and $F_Q^2 F_W$ contribute at the positions of the W atoms, and the peak strength is readily seen to be $(S_P^2 + S_Q^2) f_{Wk} = S_N^2 f_{Wk}$. These results may be compared with row 2 in Table 3.

Derivation for the noncentrosymmetric beta synthesis

We shall consider this also to indicate the method of approach in examples other than the α synthesis, which alone can be worked out exactly. When the input data are F_I, the β synthesis is clearly $|F_N|^2 \tilde{F}_I$. This can be put in the following form:

$$\frac{|F_N|^2}{\tilde{F}_I} = \frac{F_N \tilde{F}_N}{\tilde{F}_I} = \frac{F_N(\tilde{F}_I - \tilde{F}_W + \tilde{F}_Q)}{\tilde{F}_I} \quad (14a)$$

$$= F_N + \frac{(F_P + F_Q)(\tilde{F}_Q - \tilde{F}_W)}{\tilde{F}_I} \quad (14b)$$

$$= F_P + F_Q + \frac{1}{\tilde{F}_I}(F_P \tilde{F}_Q + |F_Q|^2 - F_P \tilde{F}_W - F_Q \tilde{F}_W) \quad (14c)$$

In (14c) the peaks are all known exactly, except for the contribution from the reciprocal synthesis $1/\tilde{F}_I$. This has first-order peaks at $\mathbf{r}_{Ik}$ of strength $(1/\sigma_1^2) f_{Ik}$. Using this, the results shown in row 3 of Table 3 follow.

Discussion for the limiting case when σ_1^2 tends to unity

Two limiting cases of the data given in Table 3 are of particular interest: (*a*) when there are no wrong atoms, and (*b*) when the input atoms consist of almost all the atoms in the unit cell and the wrong atoms are very few in number. The former has already been considered in Ch. 6, where it was pointed out that the β synthesis gives rise to much larger peak strengths at the unknown atomic positions, relative to the known P atoms, than the γ' synthesis.

Consider the example when the input consists of all the atoms, but a small number of these are wrong. This will correspond to the case in which $\sigma_1^2 = \chi_1^2$ and $\chi_1^2 \to 1$. The values of the enhancement factors under these limiting conditions are $\rho_Q = 1$, $\rho_W = 0$ for the noncentrosymmetric case and $\rho_Q = 0$, $\rho_W = -1$ in the centrosymmetric case. It will be seen that ρ_W is not only smaller than ρ_P, but even smaller than ρ_Q for both centrosymmetric and noncentrosymmetric crystals in the limiting case, for the β synthesis. Thus, this synthesis shows wrong atoms with negative peak strengths in centrosymmetric crystals for large enough values of σ_1^2 and greatly reduced strengths in other cases. A verification of these results is given in Ch. 8.

References

[1] A. R. Kalyanaraman, S. Parthasarathy, and G. N. Ramachandran. *The Beta Synthesis and its application to crystal structure analysis*. In "Physics of the Solid State" Eds. S. Balakrishna, M. Krishnamurthy and B. Ramachandra Rao, Academic Press (1969) London 63–76.

[2] S. Raman and W. N. Lipscomb. *Application of Fourier transform theory to electron density extraction of Patterson function.* Z. Kristallogr. **119** (1963) 30–41.

[3] M. Sax. *Convolution applied to trial and error method of crystal structure analysis.* Acta Crystallogr. **16** (1963) 439–443.

[4] R. Srinivasan. *Syntheses for the deconvolution of the Patterson function. Part IV. Refinement of the theory and a general comparison of the various syntheses.* Acta Crystallogr. **14** (1961) 607–611.

8

Tests of the various syntheses

Introduction

The results of the foregoing chapters clearly indicate that there are at least two syntheses, namely the alpha and beta syntheses, in addition to the conventional heavy-atom synthesis (γ'), which are worth detailed trial in structure determination. While the beta synthesis is the best of the three, judged purely from the point of view of its ability to reveal the unknown part of the structure with relatively much higher peak strengths compared to those of the atoms fed in, detailed tests in practical cases are obviously needed to verify this prediction of the theory. In this chapter, we shall be concerned primarily with a discussion of such tests carried out, both on hypothetical and actual examples. These can be broadly classified under two classes. The first deals with the use of these new syntheses, such as the α and the β, for regular structure determination in unknown cases, either independently, or in conjunction with the usual heavy-atom method. The second consists of cases in which specific tests have been carried out on known examples, or hypothetical models, to assess their relative merits. We shall briefly summarize the studies here and consider a detailed comparison of the results in a later section.

As mentioned in the earlier chapters, we shall concentrate our attention mainly on the three syntheses, α, β, and γ'. We have also seen that the α synthesis is nothing but a weighted sum function.[4] The various applications that have been made of the sum function[19,21,38] and also the related vector-convergence methods[3,7,8,14,30,31] may all be taken as an indication of the usefulness of the α synthesis. In its specific Fourier form, the α synthesis was tested by Raman[26] on a hypothetical point-atom model. A number of other syntheses like the β and isomorphous syntheses (see Ch. 10), were also tried by him,[26] but these were also only on hypothetical point-atom models.

A particularly detailed test of the α synthesis, especially from the point of view of actual calculation, as well as the geometrical operation of the sum function, has been made by Friedrichsons and Mathieson.[9] They have also tested on this same model the β synthesis and a good comparison of these is available in their paper. Their tests, however, are on projection data. The α synthesis, in three dimensions, has also been tested by Nowacki and Bousma.[20]

The first specific use of the beta synthesis would appear to have been made by Loopstra and MacGillavry[15] for the location of the oxygen atom in the structure of mercallite ($KHSO_4$). The usual heavy atom method was not successful in this case in revealing the oxygen positions, after the potassium and sulfur atoms were located. The β synthesis was used in the structure determination of L-cystine hydrobromide by Anantakrishnan and Srinivasan[2] in projection and by Mazumdar and Srinivasan[16–18] in three-dimensional structure determination of L-arginine hydrobromide. The β synthesis was also used by Kartha et al.[13] in the structure determination of a complex organic compound, namely that of p-bromobenzene sulfonyl ester of morellin.

Tests on the beta synthesis, particularly for assessing its relative merits with respect to the other two syntheses, namely α and γ', were carried out by Ramachandran and Ayyar,[23] Mazumdar and Srinivasan,[18] Chacko and Mazumdar,[5] and by Friedrichsons and Mathieson.[9] Detailed tests on projections of various types of functions, such as the minimum function, the α_{gen}, α_{mod}, and β_{mod} syntheses, were made earlier by Srinivasan and Aravindakshan.[36] The first group of tests [18,23,27] were made with a view to compare the β synthesis with the conventional heavy-atom synthesis. One study[9] was made with a view to assess its relative merits with respect to the α synthesis. This report also contains a critical study of the Fourier method of calculating these functions, as compared with the geometrical method of obtaining the weighted sum function.

More recently,[22] tests have been carried out to assess how the beta synthesis fares when the trial model includes a few wrong atoms. As was mentioned in the last chapter, if σ_1^2 is sufficiently large for the centrosymmetric case, the wrong atoms will be expected to occur with negative strengths. This theoretical prediction has been verified.[22] Before we consider these in detail, however, we have to examine a few practical aspects in connection with the calculations of Fourier syntheses, namely the use of weighting functions which become particularly important for the beta class of syntheses.

Weighting functions for gamma prime and beta syntheses

In the usual heavy-atom synthesis using $|F_N| \exp(i\phi_P)$ as coefficients, not all coefficients have equal importance. Simple considerations show, for example, that the phase calculated from the P atoms has a much higher probability of being correct when $|F_P|$ is large. Conversely, the errors in the phase angle ϕ_P are likely to be large when the structure amplitude $|F_P|$ is small. In order to take into account these factors, a simple procedure that may be employed is to omit those coefficients which correspond to small $|F_P|$ from the Fourier summation, by using a cutoff weighting factor based on the magnitude of $|F_P|$. A more rigorous way of handling the problem is to make use of statistical considerations. Such a study has been made by Woolfson,[39] Sim,[33] and Srinivasan,[35] and we shall briefly outline the result, following the treatment of Srinivasan.[35]

The ideal Fourier that would reveal the unknown part of the structure $(Q = N - P)$ is clearly one using $|F_Q| \exp(i\phi_Q)$ as coefficients. Since this is unknown, we first express this in the form

$$|F_Q| \exp(i\phi_Q) = |F_N| \exp(i\phi_N) - |F_P| \exp(i\phi_P). \tag{1}$$

We could now apply a well-known statistical principle which states that, in the absence of exact knowledge about a quantity (let us call it the unknown), the best procedure is to replace the unknown quantity by its mean expectation value. This is otherwise known as the principle of maximum likelihood. If there are any constraints on the unknown, these have to be taken into account. Thus, in our present case, the only unknown quantity on the right-hand side of (1) is $\exp(i\phi_N)$, since $|F_N|$ is known from the intensity of the reflection, and $|F_P|$ and $\exp(i\phi_P)$ are known from the known positions of the P atoms. Thus, replacing the unknown quantity $\exp(i\phi_N)$ on the right-hand side of (1) by its expected mean value under these conditions, we obtain

$$[F_Q] = |F_N| \langle \exp(i\phi_N) \rangle_{|F_N|, F_P} - |F_P| \exp(i\phi_P), \tag{2}$$

where $\langle \exp(i\phi_N) \rangle$ is the mean value of $\exp(i\phi_N)$, and the subscripts $|F_N|$ and F_P denote that it is the conditional mean, with these kept constant, that is to be taken. The quantity $[F_Q]$ denotes the best approximation in the above sense, which is what we seek to determine.[†] Since ϕ_P is a constant, we may write (2) as

$$[F_Q] = |F_N| \exp(i\phi_P) \langle \exp(i\phi) \rangle_{|F_N|, F_P} - |F_P| \exp(i\phi_P), \tag{3a}$$

where

$$\phi = \phi_N - \phi_P \tag{3b}$$

[†] The best Fourier synthesis defined in another context is discussed in Ch. 10. This was first discussed by Blow and Crick (Ref. 2, Ch. 10) who used the criterion that the best Fourier synthesis is the one that minimizes the sum of the squares of the densities on the difference Fourier map. The weighting function derived here can also be deduced using the criterion of Blow and Crick, as was, in fact, done by Sim for the noncentrosymmetric case.

and ϕ is shown in Fig. 1. It is clear from Fig. 1 that for fixed values of F_P and $|F_N|$ the angles $+\phi$ and $-\phi$ have both equal probability of occurrence. Thus

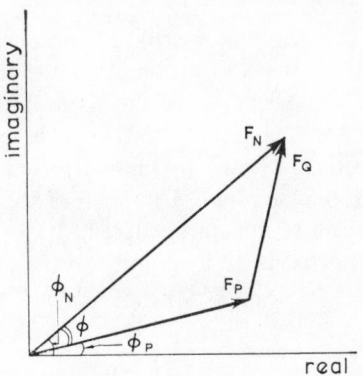

Fig. 1. Relation between the structure factors of the known part, F_P, the unknown part, F_Q, and the entire structure, F_N.

the mean value $\langle \exp(i\phi) \rangle$ can be replaced by $\langle \cos \phi \rangle$ since the mean of the sine component would vanish. Its value can be estimated knowing the probability distribution of ϕ. It turns out to be the function[33,37]

$$W_A = \langle \cos \phi \rangle = \frac{I_1(2X)}{I_0(2X)}, \tag{4}$$

where

$$X = \frac{|F_N||F_P|}{\sum_j f_{Qj}^2}, \tag{5}$$

and $I_0(X)$ and $I_1(X)$ are Bessel functions with imaginary argument of order zero and one, respectively. The symbol W_A is used to denote that it is a weighting function used to multiply the usual coefficient $|F_N| \exp(i\phi_P)$ of the γ' synthesis (the subscript A in W_A stands for the "acentric" case). The second term in (3a) may be recognized as the Fourier synthesis of the known part P which has been subtracted, so that the peak strength at the known atomic positions would be suppressed. The first term in (3a) corresponds to the conventional weighted heavy-atom synthesis.[33,39] Thus each reflection in this synthesis should be weighted by the appropriate W_A, which is calculable, since all the quantities involved in X are known. An essentially similar argument in the centrosymmetric case[34,35,39] leads to the following result for the function W_C

$$W_C = \tanh X, \tag{6}$$

where X is the same quantity as defined by (5). A basic assumption in the above argument is that ϕ_P is the "true" phase angle of the known part. In most cases in which the heavy atoms have been correctly determined, say from Patterson function, the above condition will be very well satisfied. However, if there are possible errors in the coordinates of the P atoms, the calculated values of ϕ_P will have errors in them. The probability distribution of ϕ in such a case is then affected by these errors. It has been shown that this can also be taken into account. The form of the functions $I_1(X)/I_0(X)$ and $\tanh X$ can be shown to remain unaltered,[34,37] except that the quantity X in (4) and (6) has to be replaced by a function U defined by

$$U = \frac{|F_N| \, |F_P^c| \, \sigma_A}{S_N \, S_P (1 - \sigma_A^2)}, \tag{7}$$

where F_P^c is the calculated structure factor of the P atoms with errors in them, and $S_N^2 = \sum_{j=1}^{N} f_{Nj}^2$, $S_P^2 = \sum_P f_{Pj}^2$, and σ_A is a parameter determinable[14,20] from the available data of $|F_N|$ and $|F_P^c|$. The parameter σ_A^2 reduces to the value $\sigma_1^2 (= S_P^2/S_N^2)$ when there are no errors in the positions of the P atoms, and U then becomes equal to X.

The weighted heavy-atom Fourier, $W_C |F_N| \exp(i\phi_P)$, was first tested by Woolfson[39] and was shown to improve the features of the map in that the unknown Q atoms stood out with relatively larger strength compared to the unweighted Fourier synthesis.

Weighting function for the beta synthesis

The need for a suitable weighting function is even more acute for the β synthesis than for the γ' synthesis, since $|F_P|$ occurs in the denominator in the Fourier coefficient of this synthesis. Because of this, coefficients with small values of $|F_P|$ tend to be very large. However, these are precisely the coefficients which have a large uncertainty in their phases. It is also seen from the nature of the coefficient of the β_{gen} synthesis that a weighting function which is of a higher power of the function W used for the heavy-atom synthesis will be suitable. Ramachandran and Ayyar[23] have suggested the use of W^2 as a suitable function for this purpose. A much simpler procedure would be to omit terms with $|F_P|$ less than a fraction (say one-third) of the mean value for the particular $(\sin \theta)/\lambda$. Such a cut-off procedure is decidedly arbitrary and could be resorted to only as a less preferable alternative. However, this worked reasonably well in the analysis of a fairly complicated structure.[13] A test of the effect of combining the weighting and cutoff procedures is briefly discussed in a later section of this chapter.

Tests of the beta and gamma prime syntheses

Hypothetical structures

That the β synthesis works well was first tested by Raman[26] in the case of a hypothetical point-atom structure. It was tested in an actual projection by Srinivasan and Aravindakshan.[36] A more detailed test was carried out by Ramachandran and Ayyar[23] on a known structure. They used the structure of P_4S_5, previously determined by van Houten and Wiebenga,[10] which belongs to the space group $P2_1$. Six out of the nine atoms in the asymmetric unit were assumed to be known and the beta synthesis in its general form was computed for the centrosymmetric and noncentrosymmetric projections. They are shown in Figs. 2 and 3. The syntheses were weighted by the function W^2 described in the last section. The figures also show the weighted heavy-atom synthesis $(W\gamma')$ for comparison, and as may be seen from them, the required Q atoms (marked by crosses) have come out with relatively much larger strength in the β synthesis compared with the γ' synthesis. Table 1 lists the ratio of the peak strengths for the unknown atoms in these syntheses to that in the actual Fourier F_N (Figs. 2c and 3c). The relative superiority of the beta synthesis over the γ' synthesis is readily seen from the mean enhancement factor for the Q atoms, which is nearly twice as large for the β synthesis as that for the heavy atom synthesis (γ').

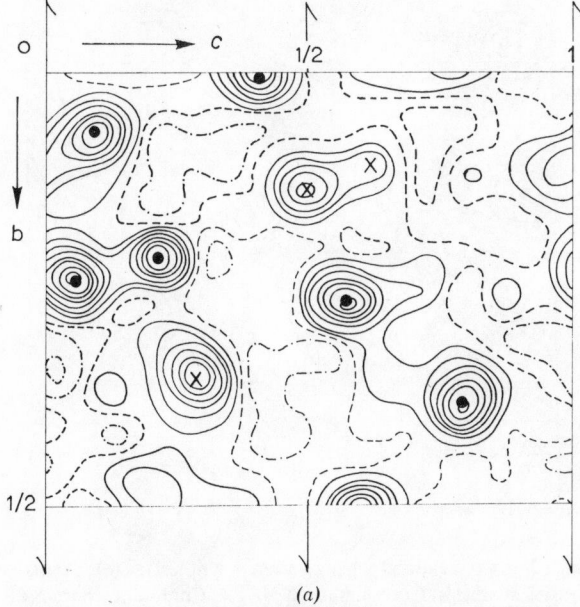

(a)

Fig. 2. Noncentrosymmetric bc projection of P_4S_5. (a) Weighted beta synthesis.

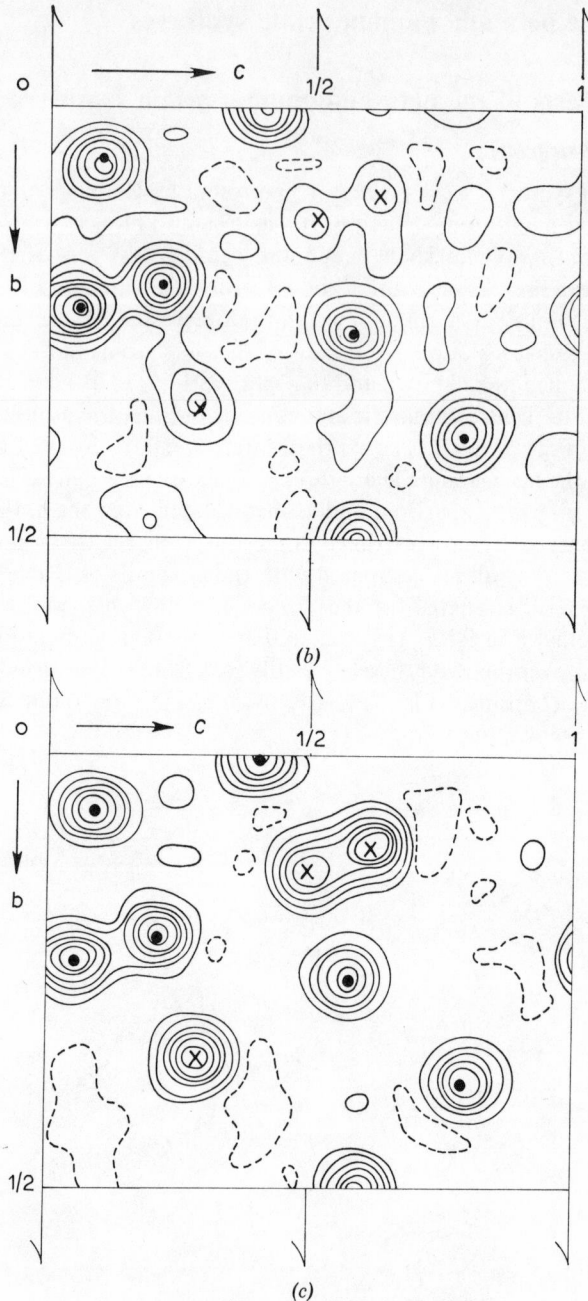

Fig. 2. (*b*) Weighted heavy-atom synthesis; (*c*) Actual Fourier synthesis. Contours are at $4e/\text{Å}^2$. Continuous lines are positive contours and dashed lines are negative contours.

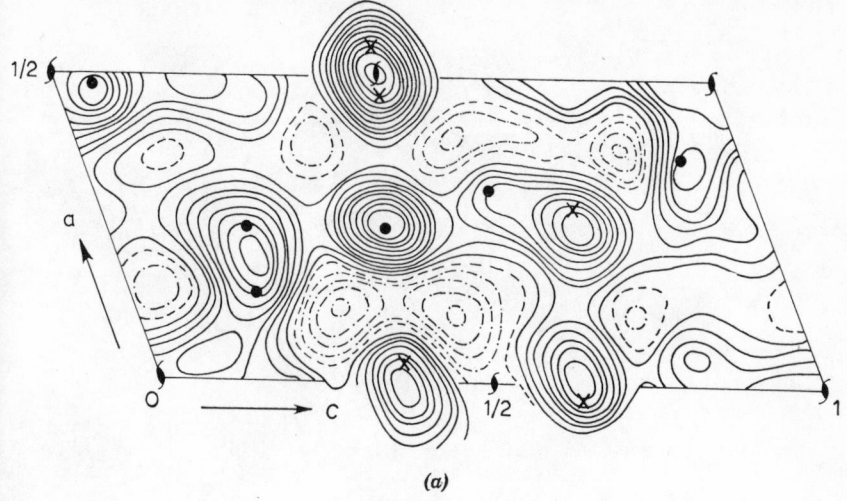

(a)

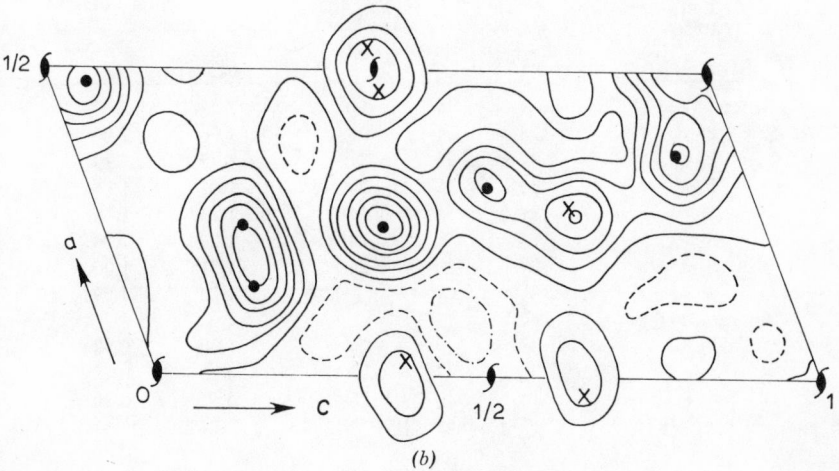

(b)

Fig. 3. Centrosymmetric ac projection of P_4S_5. (a) Weighted beta synthesis; (b) Weighted heavy-atom synthesis;

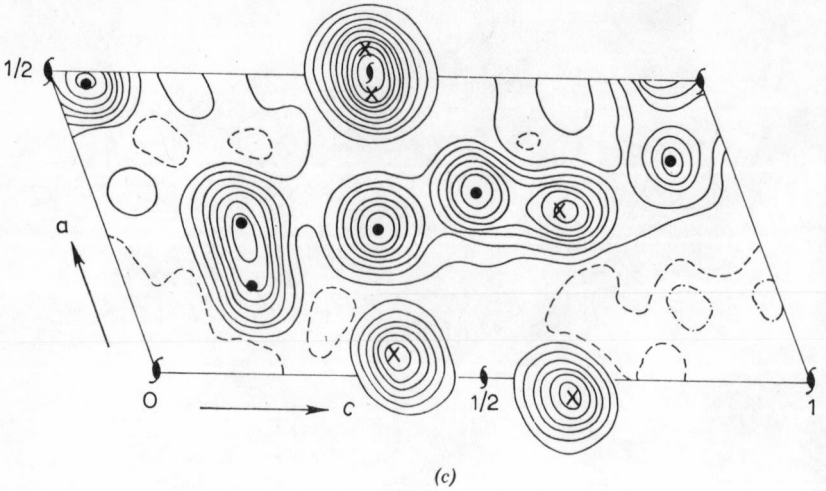

(c)

Fig. 3. (c) Actual Fourier synthesis. Contours as in Figure 2.

Table 1   Comparison of the peak strengths at the known and unknown positions in the different syntheses for a noncentrosymmetric projection (after Ramachandran and Ayyar[23])

Atom	Peak strength			Ratio to actual Fourier	
	$W^2\beta_{gen}$	$W\gamma'$	F_N	$W^2\beta_{gen}/F_N$	$W\gamma'/F_N$
Known (P)					
P_1	25	25	25	1.00	1.00
P_3	29	25	21	1.38	1.19
P_4	34	29	24	1.42	1.21
S_1	27	25	24	1.13	1.04
S_2	25	26	24	1.04	1.08
S_5	29	24	24	1.21	1.00
	Average (P)			1.20	1.09
Unknown (Q)					
P_2	24	9	24	1.00	0.38
S_3	26	13	27	0.96	0.48
S_4	16	10	28	0.57	0.36
	Average (Q)			0.84	0.41
	Ratio of Av(Q)/Av(P) (mean enhancement factor)			0.70	0.38

Actual structure analyses

Apart from the tests described above involving semihypothetical models based on known structures, instances where the β synthesis was actually used in the solution of actual structures are mentioned below. The first such instance was the case of cystine hydrobromide[2] ($C_6H_{12}N_2O_4S_2$, 2HBr). The structure belongs to the space group $P2_12_12_1$ and the bromine and sulfur atoms, determined from the Patterson function, were used for the known part P. The resultant map is shown in Fig. 4. The unknown part of the molecule is clearly discernible. The map, however, is not as good as might be expected, since the synthesis was computed without any weighting scheme, which we now know is particularly important for the β synthesis. The presence of a rather general fluctuating background and some spurious peaks in Fig. 4 may be attributed to this reason.

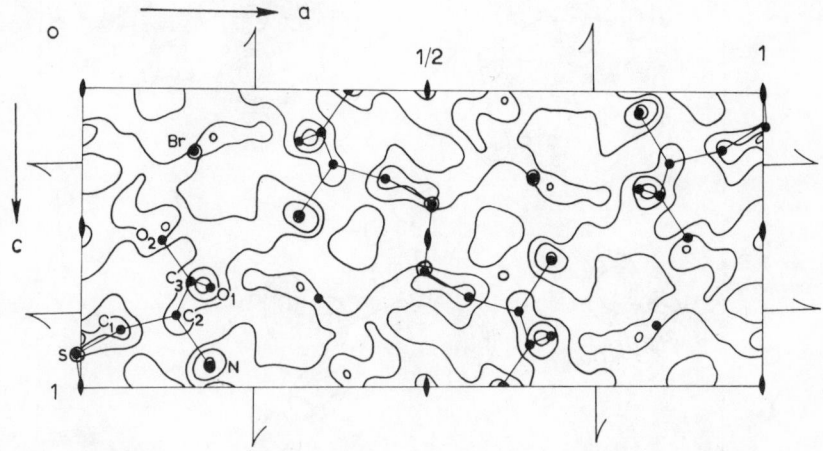

Fig. 4. Modified beta synthesis for the ac projection of L-cystine hydrobromide, using bromine and sulfur as the known part.

A more clear test in an unknown case has been made recently during the determination of the structure of L-arginine hydrobromide.[17,18] The structure belongs to the space group $P2_1$, with four molecules ($C_6H_{14}O_2N_4$, HBr, H_2O) in the unit cell. There are thus two molecules per asymmetric unit. They are approximately related by a pseudotranslational symmetry. The positions of the bromine atoms were first determined from the Patterson function and these were used to compute the β synthesis, which readily led to the structure. Since this was also a three-dimensional structure determination, it was thought worthwhile to compute the conventional heavy-atom synthesis to see how it would have fared under the same circumstances.[5] The details are presented in Fig. 5 and Table 2.

A comparison of Figs. 5a and 5b shows that the peaks in the beta synthesis are relatively much stronger than in the γ' synthesis. Table 2 lists the observed peak strengths in the two syntheses, as a ratio k, relative to the strength at the known bromine peak. The quantity k is thus proportional to the enhancement factor. The mean observed value for C, N, and O are, respectively, 0.140, 0.150, and 0.176 in the β synthesis, while they are 0.067, 0.078, and 0.087, respectively, in the γ' synthesis. This broadly agrees with the theory since, for this case ($\sigma_1^2 \simeq 0.67$), the value of k for the β synthesis is expected

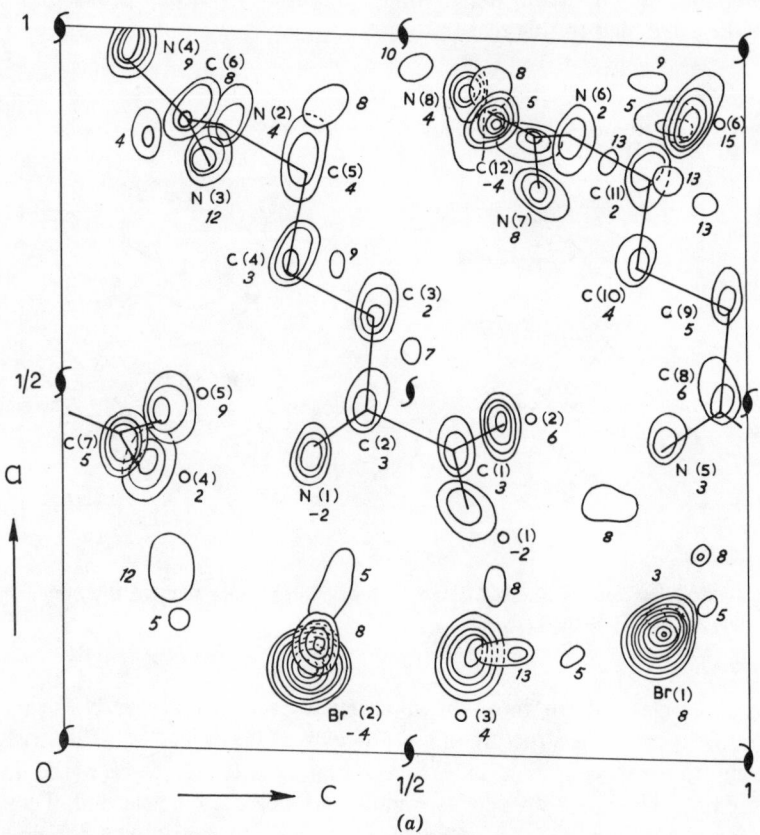

Fig. 5. Composite electron density diagram of L-arginine HBr · H$_2$O. (a) Weighted beta synthesis

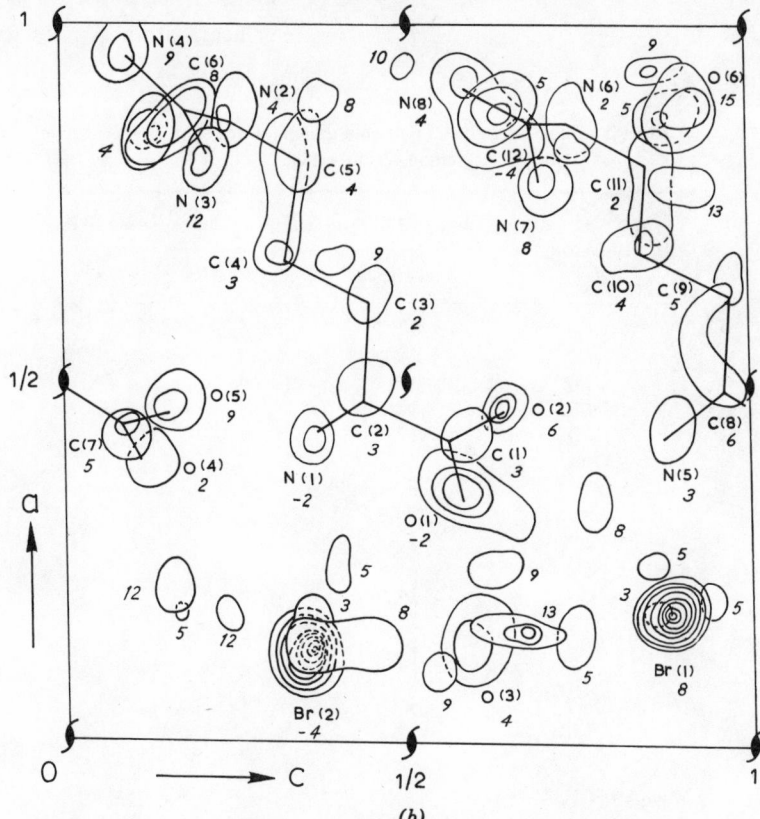

(b)

Fig. 5. (*b*) Weighted heavy-atom synthesis. Number in italics (without brackets) near each atom denotes the section (in units of 12° along *y*) in which the peak maximum occurs. The contours are at intervals of $2e/\text{Å}^3$ in both the maps. Initial contour is at $6e/\text{Å}^3$ for (*a*) and $2e/\text{Å}^3$ for (*b*). Contours for bromine alone start at $8e/\text{Å}^3$ with intervals of $8e/\text{Å}^3$ in both maps.

to be 1.7 times the value for the γ' synthesis. The comparison between theory and experiment is summarized in Table 3.

More recently, the β synthesis has been used in the structure analysis of a number of crystals in the authors' laboratory. A list of these, and the type of synthesis used, are given in Table 4. It is the general impression from these examples that the weighted β synthesis is definitely superior to the γ' synthesis in revealing the unknown atoms, and that, if the weighting is properly done, the fluctuations of its background are not inferior to those in the γ' synthesis.

Table 2   Peak strengths in beta and gamma prime syntheses of
arginine hydrobromide[5]

Atom	Value of the ratio k in per cent for	
	β synthesis	γ' synthesis
O(1)	17	8
O(2)	18	10
O(3)	19	9
O(4)	17	9
O(5)	17	9
O(6)	17	8
Mean O	17.6	8.7
N(1)	14	7
N(2)	14	8
N(3)	15	8
N(4)	19	10
N(5)	12	6
N(6)	14	7
N(7)	14	7
N(8)	17	9
Mean N	15.0	7.8
C(1)	16	8
C(2)	12	5
C(3)	15	6
C(4)	16	7
C(5)	12	5
C(6)	16	8
C(7)	14	6
C(8)	13	6
C(9)	13	6
C(10)	13	7
C(11)	12	6
C(12)	17	10
Mean C	14.0	6.7

Table 3   Comparison of the theoretical and observed values of the ratio of peak strengths k at the unknown atoms and the known bromine atom in arginine hydrobromide

Atom type	Beta		Gamma-prime	
	Observed	Theory[†]	Observed	Theory[†]
Carbon	0.140	0.117	0.067	0.037
Nitrogen	0.149	0.136	0.078	0.043
Oxygen	0.176	0.156	0.087	0.050

[†] Corresponds to $\sigma_1^2 = 0.67$ for $\theta = 0$ in the expressions (second approximation) in Table 2, Ch. 7.

Table 4   List of crystal-structure analyses in which beta syntheses were used

Crystal	Nature of synthesis used	Reference
1. L-Cystine HBr	Unweighted β synthesis, b projection	2
2. L-Arginine HBr, H_2O	Weighted β synthesis, 3D	17,18
3. Strontium tartrate, $3H_2O$	Weighted β synthesis, 3D	1
4. 3-(p-Bromophenyl)-1-nitro-2-pyrazoline	Weighted β synthesis, 3D	32
5. L-Tryptophan HBr	Weighted β synthesis, b and c projections	24
6. L-Valine HCl, H_2O	Weighted β synthesis, c projection	28
7. DL-Ornithine HBr	Weighted β synthesis, a projection	11
8. p-Bromoanil derivative of 6-methyl purine	Weighted β synthesis, a projection	12
9. D-Leu-Gly HBr	Generalized β synthesis	29

Tests of α and β syntheses

Friedrichsons and Mathieson[9] carried out tests on the centrosymmetric projection of tosyl-L-prolyl-L-hydroxyproline monohydrate ($C_{17}H_{24}O_7S$). They applied the alpha synthesis both by actual calculation of the synthesis using $|F_N|^2 F_P$ as coefficients, as well as in the form of the weighted sum function, carried out geometrically on the Patterson. The known part was taken to be the sulfur atoms to start with ($\sigma_1^2 = 0.22$), and later it was enhanced by including a few light atoms. They concluded that, purely as a geometrical operation of image seeking, the process does not materially improve the resultant map if a number of additional known light atoms are included. For instance, the first operation of inclusion of the S–S interaction revealed almost all that was desired, and inclusion of light atoms interactions, going up to as much as 50 per cent, did not improve the map to an extent one would expect from theory. Thus in such multiple superpositions using maps available on a grid, one important factor may be the possible accumulation of errors involved in the purely geometrical operation of translation and addition. In this respect, the analytical approach of direct calculation of the weighted sum function would be less susceptible to such errors. In fact this was confirmed, since the α synthesis was found to be better than the weighted sum function.

Friedrichsons and Mathieson also calculated the β synthesis for the same situations as above and concluded that it was not as good as the α synthesis for small values of σ_1^2. Their impression was that the two syntheses showed complementary properties, the beta synthesis tending to be really more efficient for larger values of σ_1^2. For instance, increased strength and sharpening were noticed in the beta synthesis. They advocated a suitable combination of the two classes of syntheses for greater efficiency. (See also a paper by Raman and Lipscomb.[27]) No quantitative data on peak strengths, however, were given by these workers. Figure 6 gives a comparison of the sum function, α synthesis and the β synthesis as reported by Friedrichsons and Mathieson. The situation represents taking as the known part the S–S interaction as well as another strong interaction in projection which corresponds to an overlap of an oxygen and a nitrogen atom.

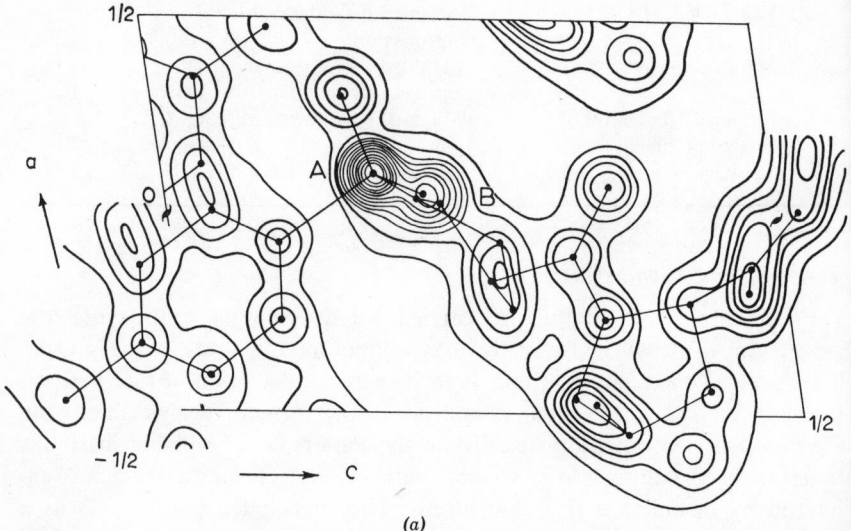

(a)

Fig. 6. Centrosymmetric ac projection of tosyl-L-prolyl-L-hydroxyproline monohydrate. (*a*) Actual Fourier synthesis;

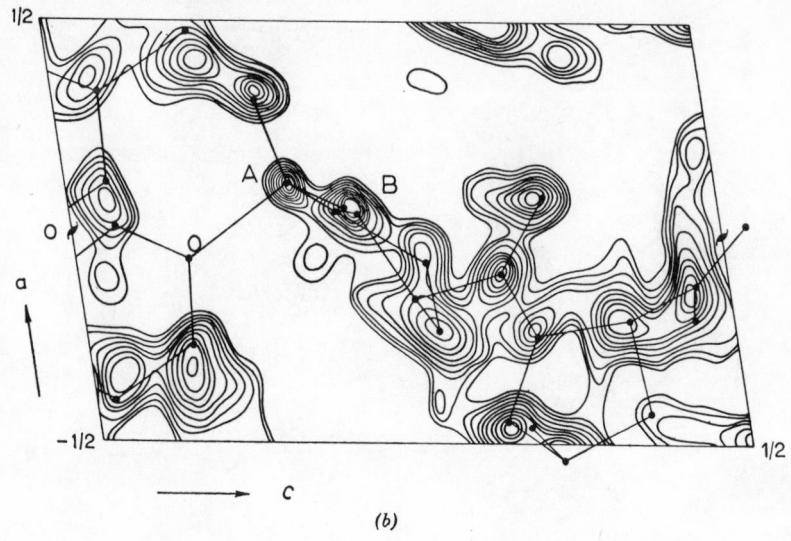

(b)

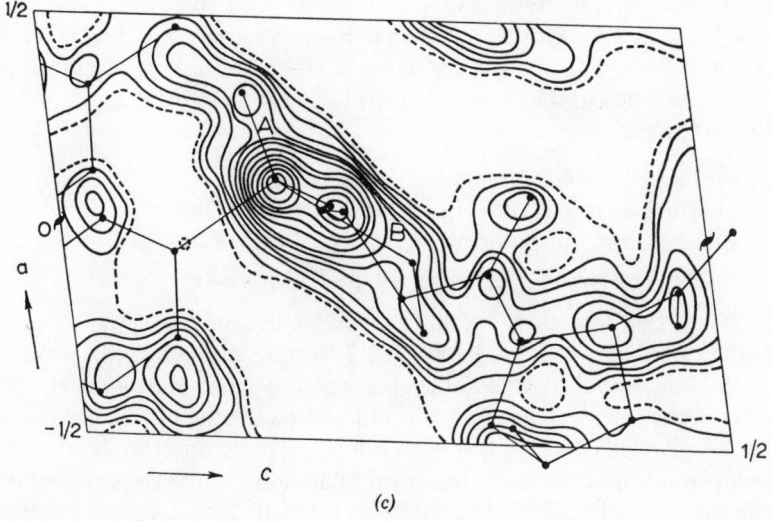

(c)

Fig. 6. (b) Sum function; (c) Alpha synthesis;

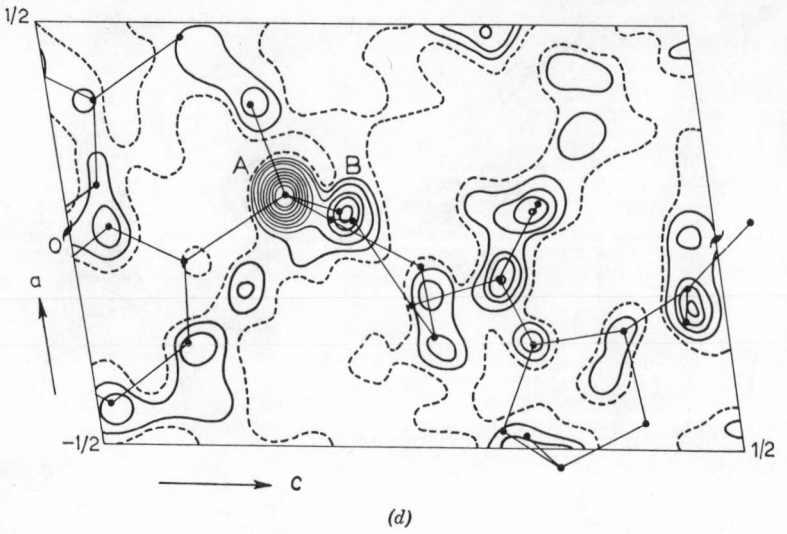

(d)

Fig. 6. (*d*) Beta synthesis with *A* and *B* as known part. *A* is a sulfur while *B*, in projection, corresponds to an overlap of an oxygen and nitrogen atom[9].

Weighting and cutoff in combination

Sometimes, even after using a weighting function W^2 for the β synthesis, a few of the terms with small $|F_P|$ may be found to contribute relatively large values to the synthesis. With a view to avoid this, the effect of an additional cutoff was tested by Parthasarathy and Kalyanaraman.[22] Using an actual crystal structure (that of 9-methyl guanine HBr), and taking a part of this ($\sigma_1^2 = 0.7$) as known, they calculated the $W^2\beta$ synthesis with the following additional procedures:

(*a*) Putting $|F_N| = 4|F_P|$, if $|F_N|/|F_P| > 4$,

(*b*) omitting all reflections with $|F_N|/|F_P| > 4$ from the synthesis

(*c*) similar to (*a*), but putting $|F_N| = 2|F_P|$, if $|F_N|/|F_P| > 2$,

(*d*) similar to (*b*), omitting all reflections with $|F_N|/|F_P| > 2$.

They found the normalized peak strengths at the unknown (Q) atoms to have the values 2.1 for (*a*), 2.1 for (*b*), 1.7 for (*c*), and 0.8 for (*d*), while the $W\gamma'$ synthesis gave a value 0.9. Thus, the relatively larger enhancement of the required atoms in the β synthesis was not affected by the procedures (*a*) and (*b*), although it led to a smoother background. Thus it appears to be advisable to use a cut-off procedure like (*a*) or (*b*) in addition to the employment of a weighting factor W^2. This has to be tested with actual cases of crystal-structure analysis.

Difference beta synthesis

A test of the difference β synthesis in an actual structure analysis has been made by Chandrasekaran.[6] It was used for the c axis projection of the centrosymmetric crystal of phenylene diamine dihydrochloride ($C_6H_8N_2$, 2HCl), space group $P\bar{1}$, with the inversion center of the molecule at the origin. The difference β synthesis was calculated using the coefficients

$$\frac{|F_0|^2}{|F_c|} \exp(i\phi_c) - |F_c| \exp(i\phi_c), \tag{8}$$

where $|F_0|^2$ were the measured (observed) intensities, and F_c were the calculated structure factors of all the atoms (excluding hydrogens). The map, shown in Fig. 7a, clearly indicated the positions of the two hydrogen atoms

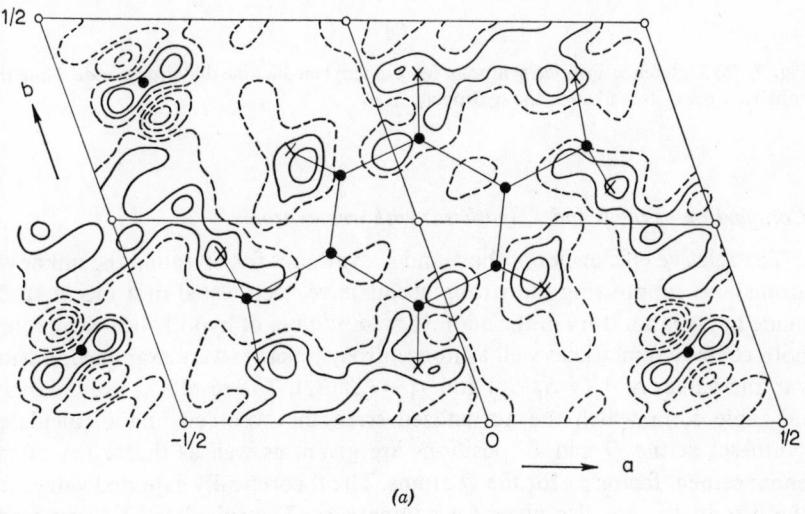

(a)

Fig. 7. (a) Difference beta synthesis;

attached to the carbon atoms of the benzene ring and one of the hydrogen atoms of the amino nitrogen. The remaining two hydrogens attached to the amino nitrogen overlapped in the projection, so that they were not separately seen, although positive peak strength was observed in the region. The difference-Fourier (difference γ') synthesis, $(F_0 - F_c)$, calculated for the same structure is shown in Fig. 7b. A comparison of the two maps shows that the difference β synthesis reveals the hydrogen atoms with nearly twice the peak heights as the difference-Fourier synthesis, as expected from theory.

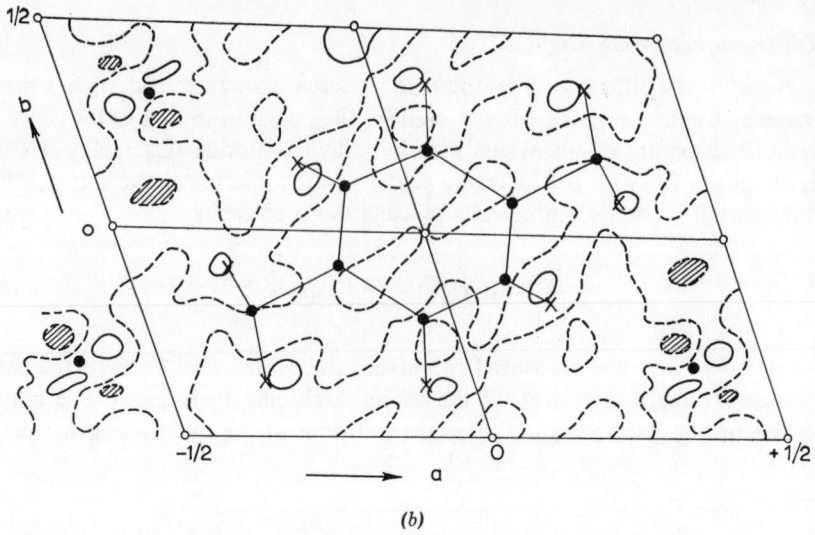

(b)

Fig. 7. (b) Difference Fourier synthesis for p-phenylenediamine dihydrochloride. Note the enhanced peak strengths for the hydrogens in a.

Comparison of the β and γ′ syntheses with wrong atoms

The relative efficiencies of the β and γ′ syntheses for revealing the unknown atoms and suppressing the wrong atoms have been tested in a recent study made in the laboratory of the authors.[22] A number of hypothetical structures, both centrosymmetric as well as noncentrosymmetric, were examined, having varying values of $\sigma_1^2 (= S_P^2/S_N^2)$ and $\chi_1^2 = (S_P^2/S_I^2)$. The results are summarized in Table 5, in which the normalized strengths observed in the calculated syntheses at the Q and W positions are given, as well as the values of the enhancement factor ρ_Q for the Q atoms. The theoretically expected values for the β synthesis are also given for comparison. The calculated Fourier maps are shown in Fig. 8. It is seen both from Fig. 8 and Table 5, that, in the non-centrosymmetric crystals, the normalized peak strength at the W atoms goes down to nearly zero as $\sigma_1^2 \to 1$ in the β synthesis, while it goes down only to a value of about 50 per cent in the γ′ synthesis. In the centrosymmetric crystals, on the other hand, the wrong atoms, in fact, come out with negative strengths in the β synthesis, as expected from theory, so that they can be immediately rejected.

The enhancement factor for the unknown Q atoms is also systematically larger for the β synthesis than the γ′ synthesis. In general, the required peaks

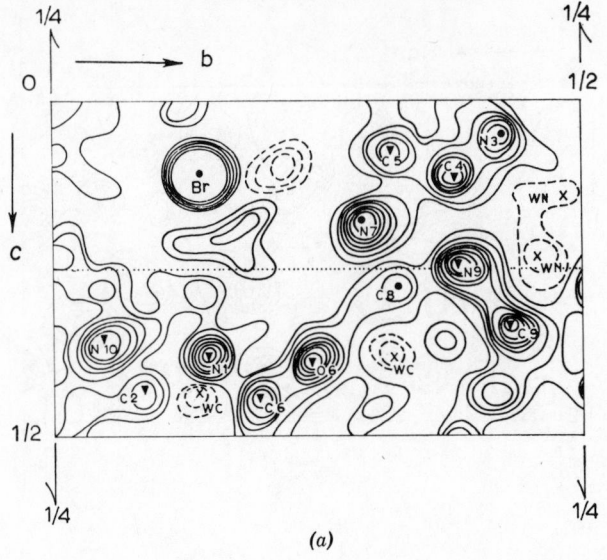

(a)

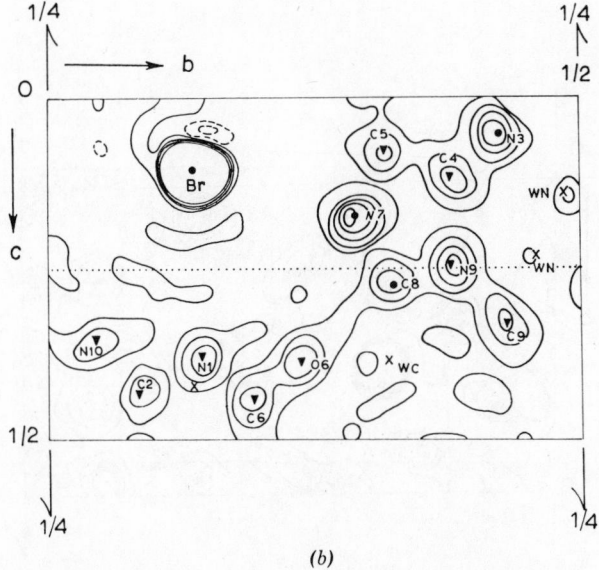

(b)

Fig. 8. Tests on the beta and heavy-atom syntheses when some of the input atoms are wrong. (a) Beta, and (b) gamma prime syntheses for the centro-symmetric case;

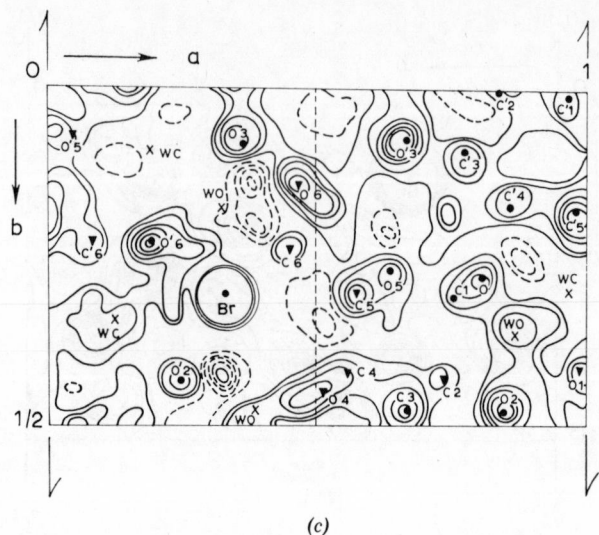

(c)

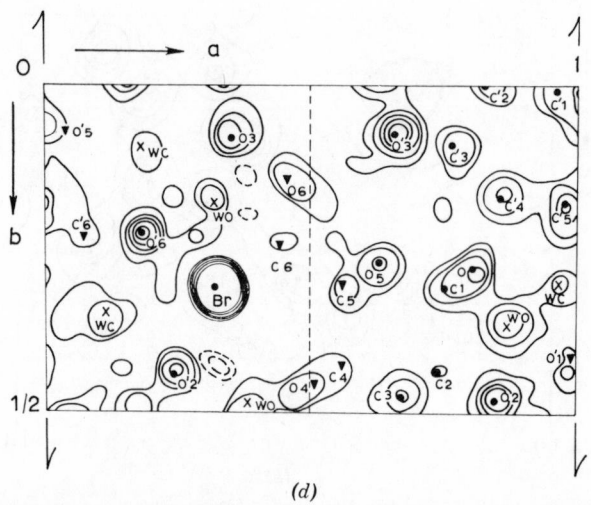

(d)

Fig. 8. (*c*) Beta and (*d*) gamma prime syntheses for the noncentrosymmetric case. Contours are at intervals of $1e/\text{Å}^2$. Starting contour is at $1e/\text{Å}^2$ and dashed lines are negative contours. (●) *P* atoms; (▲) *Q* atoms; and (*X*) = *W* atoms.

in the β synthesis are twice as strong as those in the usual heavy-atom synthesis. Thus, both as regards showing up the unknown (required) atoms and in suppressing the wrong atoms, the β synthesis is seen to be decidedly superior to the γ' synthesis.

Table 5   Observed peak strength at unknown and wrong input atoms in beta and gamma-prime syntheses with wrong atoms[†,22]

σ_1^2	χ_1^2		Mean peak strength at Q atoms		Mean peak strength at W atoms		Enhancement factor for Q atoms	
			β	γ'	β	γ'	β	γ'
			Centrosymmetric structures					
		Obs	$1.10\,f_Q$	$0.49\,f_Q$	$0.43\,f_W$	$0.70\,f_W$	0.84	0.44
0.65	0.84	Th	$0.84\,f_Q$		$0.45\,f_W$		0.65	
		Obs	$0.88\,f_Q$	$0.44\,f_Q$	$0.26\,f_W$	$0.62\,f_W$	0.76	0.43
0.82	0.87	Th	$0.87\,f_Q$		$0.19\,f_W$		0.82	
		Obs	$1.00\,f_Q$	$0.49\,f_Q$	$0.03\,f_W$	$0.52\,f_W$	0.97	0.48
0.89	0.92	Th	$0.92\,f_Q$		$0.11\,f_W$		0.89	
			Noncentrosymmetric structures					
		Obs	$1.63\,f_Q$	$0.80\,f_Q$	$-0.68\,f_W$	$0.36\,f_W$	1.42	0.87
0.78	0.90	Th	$1.80\,f_Q$		$-0.65\,f_W$		1.56	
		Obs	$1.73\,f_Q$	$0.87\,f_Q$	$-0.51\,f_W$	$0.28\,f_W$	1.44	0.83
0.78	0.89	Th	$1.78\,f_Q$		$-0.64\,f_W$		1.56	

† The theoretically calculated values for the β synthesis are also given. The theory for the γ' synthesis has not yet been fully worked out.

References

[1] G. K. Ambady. *The crystal and molecular structure of strontium tartrate trihydrate and calcium tartrate tetrahydrate.* Acta Crystallogr. **B24** (1968) 1548–1557.

[2] N. Ananthakrishnan and R. Srinivasan. *The crystal structure of L-cystine hydrobromide.* Indian J. Pure Appl. Phys. **2** (1964) 62–64.

[3] C. A. Beevers and J. H. Robertson. *Interpretation of the Patterson synthesis.* Acta Crystallogr. **3** (1950) 164.

[4] M. J. Buerger. *Vector space and its application to crystal structure investigation.* John Wiley (1954) pp. 266–268.

[5] K. K. Chacko and S. K. Mazumdar. *A comparison of the beta synthesis and the heavy atom syntheses in the determination of the structure of L-arginine monohydrobromide monohydrate.* Z. Kristallogr. **128** (1969) 315–320.

[6] R. Chandrasekharan. *The crystal structure of phenylene diamine dihydrochloride.* Acta Crystallogr. **B25** (1969) 369–374.

[7] Jerry Donohue and Kenneth N. Trueblood. *The crystal structure of hydroxy L-proline I. Interpretation of the three dimensional Patterson function.* Acta Crystallogr. **5** (1952) 414–418.

[8] Jerry Donohue and John H. Bryden. *The interpretation of the Patterson function. An application of the superposition method.* Acta Crystallogr. **8** (1955) 314–316.

[9] J. Friedrichsons and A. McL. Mathieson. *Image seeking. A brief study of its scope and comments on certain limitations.* Acta Crystallogr. **15** (1962) 1065–1074.

[10] S. Van Houten and E. H. Wiebenga. *The crystal structure of* P_4S_5. Acta Crystallogr. **10** (1957) 156–160.

[11] A. R. Kalyanaraman. *The crystal structure of DL-ornithine hydrobromide.* Curr. Sci. (India) **36** (1967) 168–169.

[12] A. R. Kalyanaraman. *The crystal structure of p-bromoanil derivative of 6-methyl purine.* In "Studies in x-ray crystallography" Ph.D. thesis (1967) University of Madras.

[13] G. Kartha, G. N. Ramachandran, H. B. Bhat, P. Madhavan Nair, V. K. V. Raghavan, and K. Venkataraman. *The constitution of morellin.* Tetrahedron letters No. 7 (1963) 459–472.

[14] W. N. Lipscomb. *Vector convergence in noncentrosymmetrical F series.* J. Chem. Phys. **26** (1957) 713–714.

[15] Lidy H. Loopstra and Caroline H. MacGillavry. *The crystal structure of* $KHSO_4$ (*Mercallite*). Acta Crystallogr. **11** (1958) 349–354.

[16] S. K. Mazumdar. *A test of the beta isomorphous synthesis.* Indian J. Pure and Appl. Phys. **3** (1965) 411–413.

[17] S. K. Mazumdar and R. Srinivasan. *X-ray analysis of L-arginine hydrohalides.* Curr. Sci. (India) **33** (1964) 573–575.

[18] S. K. Mazumdar and R. Srinivasan. *The crystal structure of L-arginine mono-hydrobromide monohydrate.* Z. Kristallogr. **123** (1966) 186–205.

[19] Dan McLachlan Jr. *The determination of crystal structures from x-ray data without a knowledge of the Fourier coefficients.* Proc. Nat. Acad. Sci. (U.S.) **37** (1951) 115–124.

[20] von W. Nowacki and G. F. Bousma. *Die Kristall and molekulstruktur von Erythralin-hydrobromid* $C_{18}H_{19}O_3N \cdot HBr$ Z. Kristallogr. **110** (1958) 89–111.

[21] William Revel Philips and Dan McLachlan Jr. *A versatile projection for assisting in crystal structure determination.* Rev. Sci. Inst. **25** (1954) 123–128.

[22] A. R. Kalyanaraman, S. Parthasarathy and G. N. Ramachandran. *The Beta synthesis and its application to crystal structure analysis.* In "Physics of the Solid State". Eds. S. Balakrishna, M. Krishnamurthy and B. Ramachandra Rao. Academic Press, London (1969) 63–76.

[23] G. N. Ramachandran and R. R. Ayyar. *Fourier syntheses for feeding in isomorphous replacement and anomalous dispersion data.* In "Crystallography and crystal perfection" Ed. G. N. Ramachandran (1963) Academic Press, London, 25–41.

[24] R. Ramachandra Iyer and R. Chandrasekharan. *The crystal structure of L-tryptophan hydrobromide.* Curr. Sci. (India) **36** (1967) 139–143.

²⁵ R. Ramachandra Ayyar and R. Srinivasan. *The crystal structure of* L(+)*cysteine hydrochloride monohydrate.* Curr. Sci. (India) **34** (1965) 449–450.

²⁶ S. Raman. *Syntheses for the deconvolution of the Patterson function. Part II. Detailed theory for non-centrosymmetric crystals.* Acta Crystallogr. **12** (1959) 964–975.

²⁷ S. Raman and W. N. Lipscomb. *Application of Fourier transform theory to electron density extraction of the Patterson function.* Z. Kristallogr. **119** (1963) 30–41.

²⁸ S. T. Rao. *Crystal structure of* L-*valine hydrochloride monohydrate.* In "Structure studies on organic compounds". Ph.D. thesis (1966). University of Madras.

²⁹ S. T. Rao. *Crystal structure of* D-*leucyl glycine hydrobromide.* In "Structure studies on organic compounds" Ph.D. thesis (1966) University of Madras.

³⁰ J. H. Robertson. *Interpretation of the Patterson synthesis. Rubidium benzyl penicillin.* Acta Crystallogr. **4** (1951) 63–66.

³¹ J. H. Robertson and C. A. Beevers. *The crystal structure of strychnine hydrogen bromide.* Acta Crystallogr. **4** (1951) 270–275.

³² M. N. Sabesan and K. Venkatesan. *The crystal structure of* 3-(*p-bromophenyl*)-1-*nitro-2-pyrazoline.* Private communication.

³³ G. A. Sim. *A note on the heavy atom method.* Acta Crystallogr. **13** (1960) 511–512.

³⁴ R. Srinivasan. *Weighting functions for use in the early stages of structure analysis when a part of the structure is known.* Acta Crystallogr. **20** (1966) 143–144.

³⁵ R. Srinivasan. *Weighting function in crystal structure analysis.* Z. Kristallogr. **126** (1968) 175–181

³⁶ R. Srinivasan and C. Aravindakshan. *Syntheses for the deconvolution of the Patterson function. Part V. Test of the various syntheses for centrosymmetric crystals.* Acta Cryst. **14** (1961) 612–616.

³⁷ R. Srinivasan and R. Chandrasekharan. *Correlation function connected with structure factors and their application to observed and calculated structure factors.* Indian J. Pure Appl. Phys. **4** (1966) 178–186.

³⁸ Ivor D. Thomas and Dan McLachlan Jr. *Some critical tests of the use of mixed projections in crystal structure determination.* Acta Crystallogr. **5** (1952) 301–306.

³⁹ M. M. Woolfson. *An improvement of the heavy-atom method of solving crystal structures.* Acta Crystallogr. **9** (1956) 804–810

9

Isomorphous replacement: general aspects

Introduction

Next to the heavy-atom method, the technique of isomorphous replacement is one of the earliest and most commonly used ones in crystal-structure analysis. The applications of this method to the complete determination of signs of structure factors in centrosymmetric crystals[7,8,17] and to the determination of phase angles in noncentrosymmetric crystals[1-3,9] are now well known.[†] In this chapter, we shall apply the ideas developed in this monograph which enable us to obtain various types of Fourier syntheses, making use of data from isomorphous crystals to yield the unknown part of the structure. We shall first outline briefly the classical methods of sign and phase determination, employing isomorphous-crystal data.

Suppose that, out of a total number (N) of atoms in a crystal structure, a group of P atoms is replaced by another set of P atoms with different scattering power. We shall assume that there is no shift in the positions of both the P atoms and the remaining Q atoms, except for the change in scattering power of the P atoms. Let us denote the scattering power of the P atoms in the two crystals by f_P^{I} and f_P^{II} and we shall assume for convenience that $f_P^{II} > f_P^{I}$. We will be using the superscripts I and II to denote all the quantities relevant to crystals I and II, respectively. The above type of situation will be referred to as "isomorphous replacement." There is another possible variation of this, namely when a few atoms, say P atoms, are *added* to the first crystal of Q atoms to form the second crystal, both crystals, however, still

[†] For a comprehensive list of references to the literature, see Ref. 5.

120

retaining the features of isomorphism, namely that the common Q atoms are in the same locations in both the crystals. This type of isomorphism is commonly met with in protein crystallography, where a few heavy atoms can be attached to a parent protein without any significant change in the protein molecule itself. In order to distinguish this type from the replacement type discussed earlier, we shall call this as "isomorphous *addition*." It may be noticed that the case of isomorphous addition is equivalent to that of replacement, when $f_P^I = 0$ in our notation. In our discussions below, we shall be dealing mainly with the general case of the replacement type unless specifically mentioned otherwise.

Phase determination with isomorphous crystals[1–4,6,9–11,13]

Figure 1 shows the relations between the relevant vectors in the Argand diagram. F_Q is the contribution to the structure factor from the common Q atoms in the two crystals. To this are added F_P^I and F_P^{II} to give, respectively, F_N^I and F_N^{II}. The difference between these is denoted by $\delta F_P \equiv (F_P^{II} - F_P^I)$. We shall use the symbol δ generally as a prefix to indicate the difference between a pair of corresponding quantities for the two isomorphous crystals, with the quantity for crystal I being subtracted from that for crystal II.

The quantity F_Q is generally the sum of a number of contributions from a large number of atoms, but F_P is usually due to a small number of heavy atoms and therefore their locations may be determined, e.g., from a Patterson synthesis. Later in this chapter and in the next, we shall consider methods of determining the positions of the replaceable P atoms even when they are not heavy. For the present discussion, we assume that the nature and locations of the replaceable atoms are known, so that it is possible to calculate F_P^I and F_P^{II}, and hence δF_P, in magnitude and direction in the Argand diagram. The quantities $|F_N^I|$ and $|F_N^{II}|$ are known from the measured intensities of the two crystals, and these magnitudes are indicated in Fig. 1 by the two circles. A solution of the problem under these conditions is shown by the vectors in the upper half of Fig. 1 (solid lines). Denoting by ϕ_P the phase of F_P (and of δF_P), the line OP is drawn at this angle to the real axis through the origin. We define

$$\theta_Q = \theta = \phi_Q - \phi_P, \tag{1a}$$

$$\theta^I = \phi_N^I - \phi_P, \qquad \theta^{II} = \phi_N^{II} - \phi_P. \tag{1b}$$

Then, taking the triangle OAB defined by the vectors F_N^I, F_N^{II}, and δF_P, it is readily seen that

$$|F_N^{II}|^2 = |F_N^I|^2 + |\delta F_P|^2 + 2|F_N^I||\delta F_P| \cos \theta^I, \tag{2}$$

which yields

$$\cos \theta^I = \frac{|F_N^{II}|^2 - |F_N^I|^2 - |\delta F_P|^2}{2|F_N^I||\delta F_P|}. \tag{3}$$

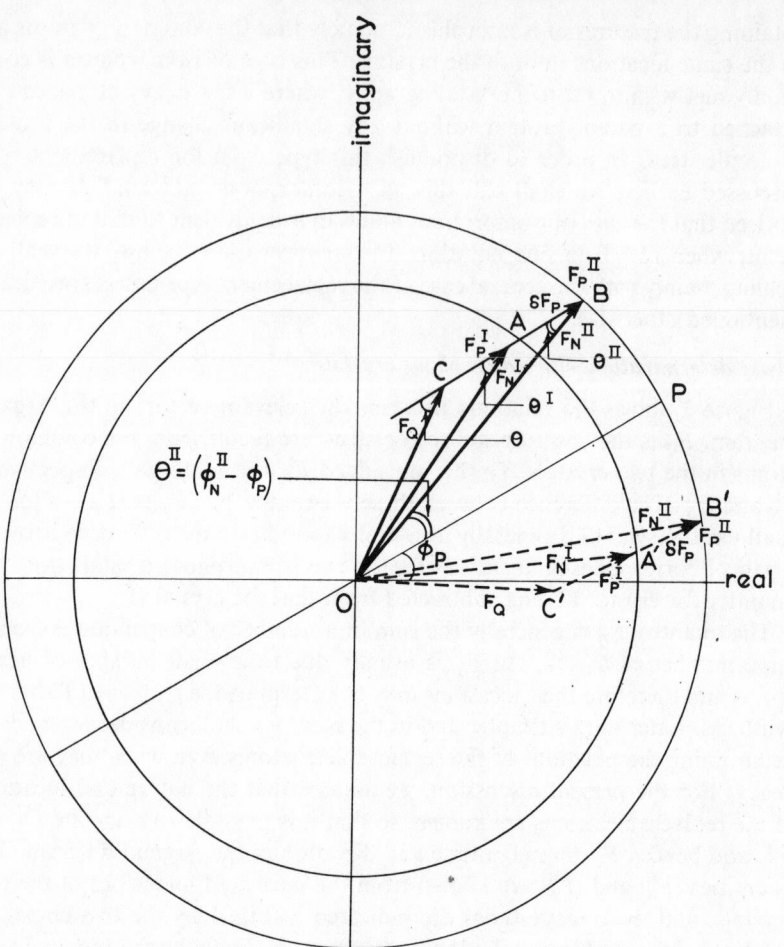

Fig. 1. Phase determination with isomorphous data. Note the symmetry of the two ambiguous solutions with respect to the direction of the vector F_P.

Since this is a relation involving the cosine of the angle θ^I, two possible solutions exist for θ^I, namely $+\theta^I$ and $-\theta^I$, given the magnitude of the right-hand side of (3). The second solution $-\theta^I$ is readily seen from Fig. 1 to be the one corresponding to a mirror image of the triangle OAB with respect to the line OP, which yields the triangle $OA'B'$. Thus the deviation of the phase angle of the structure factor F_N^I from the phase of the contribution from the replaceable atoms has an ambiguity in its sign. If the replaceable atoms are centrosymmetric, the quantities F_P^I, F_P^{II}, and δF_P become real and this is equivalent to taking the line OP to the same as the real axis, in Fig.

1. The ambiguity in phase is now reduced to an ambiguity in sign of the actual phase angle ϕ_N^{I}, i.e., $\phi_N^{I} = \pm \theta_N^{I}$.

It is clear from Fig. 1 that similar ambiguity exists for the phases of F_N^{II} and F_Q. In particular, if we consider the latter, which corresponds to the nonreplaceable and unknown part of the structure, it is seen from the triangles OAC and $OA'C'$ that its phase angle ϕ_Q has the ambiguity of a sign, leading to two values

$$\phi_Q = \phi_P \pm \theta_Q, \tag{4}$$

exactly as in the case of ϕ_N^{I}, $= \phi_P \pm \theta^{I}$. If we call one of the phases obtained, which is correct, by say ϕ_Q, the other phase is thus given by $(2\phi_P - \phi_Q)$. A similar result holds for the phase angles ϕ_N^{I} and ϕ_N^{II}—that is, the two pairs are of the form ϕ_N^{I}, $2\phi_P - \phi_N^{I}$ and ϕ_N^{II}, $2\phi_P - \phi_N^{II}$, respectively.

In an actual structure determination using the isomorphous method under these condtions, there is no way of distinguishing one of the solutions from the other as being the correct phase. The best that can be done under such circumstances is to use both the possible solutions for the phase angle in a Fourier synthesis.[1-3] We may call this as the double-phased synthesis. This can be written, for the structure F_N^{II}, as

$$|F_N^{II}| \{ \exp [i(\phi_P + \theta^{II})] + \exp [i(\phi_P - \theta^{II})] \} \tag{5a}$$

$$= |F_N^{II}| \exp (i\phi_N^{II}) + |F_N^{II}| \exp (-i\phi_N^{II}) \exp (2i\phi_P). \tag{5b}$$

Equation (5b) follows from the fact that the two phase angles for F_N^{II} are given by ϕ_N^{II} and $2\phi_P - \phi_N^{II}$, mentioned earlier.

If we consider the structure F_Q instead of F_N^{II}, the double-phased synthesis is equivalent to

$$|F_Q| \{ \exp [i(\phi_P + \theta)] + \exp [i(\phi_P - \theta)] \} \tag{6a}$$

$$= |F_Q| \exp (i\phi_Q) + |F_Q| \exp (-i\phi_Q) \exp (2i\phi_P). \tag{6b}$$

The expressions (5b) and (6b) are seen to be, respectively, equivalent to

$$F_N^{II} + \tilde{F}_N^{II} \exp (2i\phi_P) \tag{7}$$

and

$$F_Q + \tilde{F}_Q \exp (2i\phi_P). \tag{8}$$

The double-phased synthesis for the first case can therefore be interpreted as being equivalent to the structure F_N^{II} plus the inverse structure convolved with the phase-squared structure of the P atoms. Thus, basically, it contains the structure given by the first term in (7). The second term is expected only to contribute to the general background in the map, since it is a convolution of the two structures F_N^{II} and $\exp (2i\phi_P)$. An exactly similar argument holds for the case of the Fourier synthesis with coefficients represented by (8). In this case, only the Q atoms are expected to come out prominently, with a background arising from the convolution of the two structures in the second term.

P group centrosymmetric

The case when the replaceable atoms have a centrosymmetric configuration in a noncentrosymmetric crystal is of particular interest. This occurs, for example, when the number of replaceable atoms in the unit cell is just one or two, or when these atoms occupy special positions in a noncentrosymmetric space group. In such a case, ϕ_P is either 0 or π, and $\exp(i2\phi_P) = 1$ for either value. Expression (5) then becomes

$$|F_N^{II}| \exp(i\phi_N^{II}) + |F_N^{II}| \exp(-i\phi_N^{II}),$$

which is equivalent to

$$F_N^{II} + \tilde{F}_N^{II}. \tag{9}$$

(This may also be seen from the expression (7), putting $\phi_P = 0$ or π). Correspondingly, the double-phased synthesis, represented by expression (6), also becomes

$$F_Q + \tilde{F}_Q. \tag{10}$$

In either case, it follows that the structures F_N and F_Q will be duplicated by a set of peaks corresponding to their inverse structures $\tilde{F}_N$ and $\tilde{F}_Q$, respectively, the center of inversion being that of the P group of atoms. The duplicated inverse peaks will have exactly the same strengths as those in the true structure, and a distinction between these will not be possible, so that the seeking out of the correct structure from the diagram of the double-phased synthesis will present considerable difficulty in this case.

It is interesting that the case in which the P group of atoms is also noncentrosymmetric is superior in revealing the unknown Q atoms. This is due to the fact that only the true peaks occur strongly, since the inverse peaks get dispersed because the structure $\tilde{F}_N$, or $\tilde{F}_Q$ (as the case may be), is convolved with the phase-squared structure of the P atoms in this case.

Centrosymmetric crystals

When the crystal as a whole is centrosymmetric, the phase ambiguity vanishes completely. It is easiest to see this from relation (7) or (8). Since $\phi_P = 0$ or π, we get, in this case, the structures $F_N^{II} + \tilde{F}_N^{II}$ or $F_Q + \tilde{F}_Q$. The fact that the crystal as a whole is centrosymmetric means that $F_N^{II} \equiv \tilde{F}_N^{II}$ and $F_Q \equiv \tilde{F}_Q$. Thus the inverse duplicated structure mentioned in the last subsection automatically coincides with the correct structure and hence the solution becomes unique.

It is useful to consider this from another point of view. Considering Fig. 1, all the vectors F_N^I, F_N^{II}, F_Q, F_P^I, and F_P^{II} become real and lie along the real axis of the Argand diagram. Further, in (3), the angle θ^{II} can take only two values, 0 or π. Thus that equation takes the form

$$\pm 1 = \frac{|F_N^{II}|^2 - |F_N^I|^2 - |\delta F_P|^2}{2|F_N^I||\delta F_P|}, \tag{11}$$

where the plus and minus signs of ± 1 mean that F_N^{II} and F_P have the same or opposite signs. On reduction, (11) is equivalent to

$$|F_N^{II}| \mp |F_N^I| = |\delta F_P|. \tag{12}$$

This amounts to saying that the criterion, whether the difference or the sum, of the magnitudes of F_N^{II} and F_N^I is equal to δF_P, fixes whether F_N^{II} and F_P are of the same or opposite signs. Since from the known positions of the P atoms the sign of δF_P is also known, this also fixes uniquely the sign of F_N^{II}. This is the basis of the unique assignment of signs to structure factors in isomorphous centrosymmetric crystals, first employed by Robertson and Woodward.[17]

Essentially the same principles are involved in the graphical method of Hargreaves[7,8] which takes into account the possibility of difference in scale factors for the two amplitudes $|F_N^{II}|$ and $|F_N^I|$. It is readily shown that if K_1 and K_2 are the correct scale factors for the unscaled structure amplitudes $|F_N^{0I}|$ and $|F_N^{0II}|$ respectively, relation (13), below, holds good:

$$\frac{K_2 F_N^{0II}}{\delta F_P} = K_1 \frac{F_N^{0I}}{\delta F_P} + \exp\left(-\frac{B \sin^2 \theta}{\lambda^2}\right), \tag{13}$$

where

$$\delta F_P = \sum_{j=1}^{P} (f_P^{II} - f_P^I) \cos(2\pi \mathbf{H} \cdot \mathbf{r}_{Pj}), \tag{14}$$

and B is the temperature factor, assumed to be the same for both the crystals. In a narrow region of $\sin \theta$ within which B in (13) can be assumed to be a constant, a plot of $F_N^{II}/\delta F_P$ versus $F_N^I/\delta F_P$ should be a straight line. To start with a plot of $|F_N^{II}/F_P|$ versus $|F_N^I/F_P|$ could be made. Only those reflections for which F_N^{II} and F_N^I have both the same sign of δF_P will fall on the true line given by (13). For other reflections a proper assignment of sign for F_N^{II} and F_N^I relative to the sign of δF_P will make them fall on this straight-line graph. Since the sign of δF_P is known the actual signs of F_N^I and F_N^{II} could be deduced uniquely. Analogous methods, which are only extensions of the above method, have been devised, for the case of anomalous scattering data collected at two wavelengths, which yield essentially isomorphous information arising from the variation of the real correction $\delta f'$ with wavelength (see Ref. 16, Ch. 13).

The method of double isomorphous replacement[1,9,13]

While the problem of sign determination is unique for a pair of centrosymmetric isomorphous crystals, we have seen that, for a pair of noncentrosymmetric isomorphous crystals, the phase solution has an inherent 2-fold ambiguity. It was recognized by Bokhoven, Schoone, and Bijvoet[3] that the ambiguity can be resolved if an additional isomorphous compound becomes available—that is, if we have at least three compounds which are isomorphous to one another. The possible unique solution in such a case of "double isomorphous replacement" and its further implications were thoroughly examined by Harker.[9] We shall outline this method here briefly.

For convenience of discussion, we shall consider isomorphism of the addition type. Suppose we have a given crystal, say I, composed of Q atoms, which is isomorphous with two other crystals, II and III, obtained by addition of atoms, say P_2 and P_3, respectively. In our notation, this is equivalent to taking $F_P^I = 0$ in crystal I, so that

$$F^I = F_Q,$$ (15a)

$$F^{II} = F_Q + F_P^{II},$$ (15b)

$$F^{III} = F_Q + F_P^{III}.$$ (15c)

We assume that all the magnitudes $|F^I|$, $|F^{II}|$, $|F^{III}|$ and the vectors F_P^{II}, F_P^{III} are known. The theory of the isomorphous method for a pair of crystals, shown in Fig. 1, can be looked at in a slightly different manner, as in Fig. 2. Thus, we first set up the vector $-F_P^{II}$ and a circle of radius $|F^{II}|$ with the terminus of $-F_P^{II}$ (C_2) as center. A circle is drawn with the origin as center having a radius $|F^I|$, and the two circles intersect at two points K, K' (in general), which correspond to the two possible phase angles for F^I, as determined from the pair of crystals I and II. Suppose now we construct a similar circle using data of the third crystal. Thus a circle is drawn of radius $|F^{III}|$, with its center C_3 at the terminus of the vector $-F_P^{III}$. It is obvious that there will be two intersections (K, K'') of the third circle, leading to two solutions for the phase angle. However, there will be only one common intersection (K) between circles (I), (II), and (III) (when the data are all exact), and this fixes the phase angle uniquely.

It is obvious that the above procedure will lead to a unique solution only if the vectors F_P^I and F_P^{II} are not collinear (i.e., either parallel or antiparallel). This condition will be satisfied, in general, if the group of atoms P^{II} and P^{III} are acentric in configuration, and take up different positions in crystals II and III relative to the atoms of I. A few accidental cases may, however, arise when the vectors F_P^{II} and F_P^{III} are nearly collinear, in which case the two solutions are the same for both the pairs of crystals. However, these are not likely

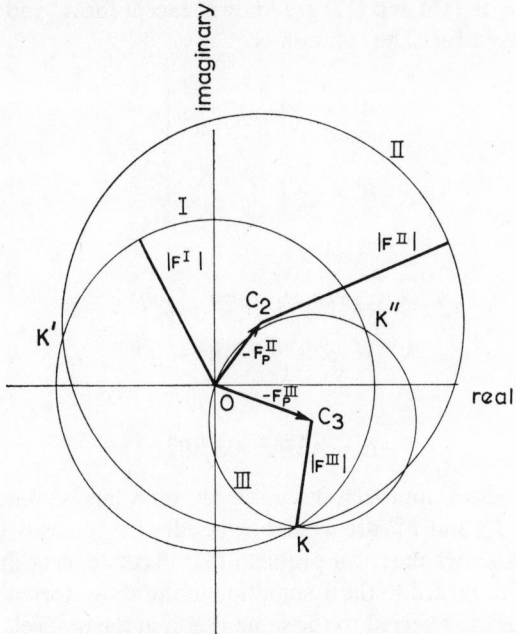

Fig. 2. Resolution of phase ambiguity with double isomorphous replacement.

to affect the solution of the structure as a whole. If the P group of atoms has a centrosymmetric configuration, then the sets P^{II} and P^{III} should not be centrosymmetric about the same point in the unit cell containing the structure F_Q, nor should they be centrosymmetric about points which are separated by half a repeat spacing of the unit cell. A case of particular interest is when P^{II} or P^{III} happen to be single heavy atoms, when they are necessarily centrosymmetric.

The method outlined above is a graphical one and can be applied in practice. An analytical approach is also possible; this corresponds to solving for the real and imaginary parts of $F^I = A^I + iB^I$ from a pair of simultaneous equations. Thus, in an obvious notation, we have

$$
\begin{aligned}
|F^{II}|^2 &= (A^I + A_P^{II})^2 + (B^I + B_P^{II})^2 \\
&= (A^I)^2 + (B^I)^2 + (A_P^{II})^2 + (B_P^{II})^2 + 2A^I A_P^{II} + 2B^I B_P^{II} \\
&= |F^I|^2 + |F_P^{II}|^2 + 2A^I A_P^{II} + 2B^I B_P^{II}.
\end{aligned}
\tag{16}
$$

Similarly

$$
|F^{III}|^2 = |F^I|^2 + |F_P^{III}|^2 + 2A^I A_P^{III} + 2B^I B_P^{III}.
\tag{17}
$$

All the quantities in (16) and (17) are known, except for A^I and B^I, and hence these can be solved for. The solution is

$$A^I = \frac{1}{2\Delta} \begin{vmatrix} K^{II} & B_P^{II} \\ K^{III} & B_P^{III} \end{vmatrix}, \tag{18a}$$

$$B^I = \frac{1}{2\Delta} \begin{vmatrix} A_P^{II} & K^{II} \\ A_P^{III} & K^{III} \end{vmatrix}, \tag{18b}$$

where

$$K^{II} = |F^{II}|^2 - |F_P^{II}|^2 - |F^I|^2 \tag{19a}$$

$$K^{III} = |F^{III}|^2 - |F_P^{III}|^2 - |F^I|^2 \tag{19b}$$

and

$$\Delta = A_P^{II} B_P^{III} - A_P^{III} B_P^{II}. \tag{20}$$

In both methods of approach discussed above, it has been tacitly assumed that the vectors F_P^{II} and F_P^{III} are available, besides the intensity data from the three crystals. Another practical problem that occurs concerning the vectors F_P^{II} and F_P^{III} is with regard to the assumption in the above formulas that all the three compounds are referred to the same origin in the unit cell in calculating the various structure factors. This may not be possible to achieve without further effort in all the space groups. For instance, in the space group P2$_1$, the choice of origin is arbitrary along the 2-fold screw axis. Thus when the position of the heavy atoms P^{II} and P^{III} have been determined individually, say from their Patterson syntheses $|F^{II}|^2$ and $|F^{III}|^2$, it is still necessary to ensure that they are referred to the same origin of reference on the screw axis. This is equivalent to the problem of obtaining the relative positions of the heavy atoms in the two crystals with reference to the first parent crystal, F^I. This could be done in the analytical approach by the method of successive approximations. A better method is to use special correlation Patterson-type syntheses, making use of the intensity data from the two crystals II and III. We shall consider these in Chapter 10 as part of the methods for the location of heavy atoms in isomorphous crystals.

Simulation of isomorphism by use of different wavelengths[11,12,14-16]

We have so far discussed some of the well-known methods of treating the data obtained from more than one crystal which are structurally isomorphous in the chemical sense. The phenomenon of isomorphism can be simulated (as far as the intensity data are concerned) with a single crystal by making use of the anomalous dispersion effect which was briefly referred to in (15),

Ch. 1. Thus, even if we have only one crystal, if the structure contains a suitable anomalous scatterer whose scattering factor is therefore dependent on the wavelength, we can choose two different wavelengths for which the corrections $\delta f'$ for the anomalous scatterer are different. The intensity data will therefore be different for the two wavelengths, and it is obvious that, since it is only the real component of the correction that changes, this device is essentially equivalent to simulating isomorphous data. In principle, the various results discussed above, and also those of Ch. 10, are equally applicable to such data. In fact, some of the practical difficulties, which are met with in the case of a pair of different crystals such as lack of perfect isomorphism, are absent in this case, and the isomorphism is exact, since there is no change in the coordinates of the anomalous scatterers. We shall have occasion to discuss anomalous dispersion effects in greater detail in Ch. 11–13.

References

[1] J. M. Bijvoet. *Phase determination in direct Fourier synthesis of crystal structures.* Proc. Konikl. Ned. Akad. Wetenschap. **52** (1949) 313–314.

[2] C. Bokhoven, J. C. Schoone, and J. M. Bijvoet. *On the crystal structure of strychnine sulphate and selenate III.* (0 0 1) *projection.* Proc. Koninkl. Ned. Akad Wetenschap. **52** (1949) 120–121.

[3] C. Bokhoven, J. C. Schoone, and J. M. Bijvoet. *The Fourier synthesis of the crystal structure of strychnine sulphate pentahydrate.* Acta Crystallogr. **4** (1951) 275–280.

[4] D. M. Blow and M. G. Rossmann. *Single isomorphous replacement method.* Acta Crystallogr. **14** (1961) 1195–1202.

[5] M. J. Buerger. *Crystal-structure analysis.* (Wiley, New York, 1960) pp. 530–532.

[6] J. M. Cork. *The crystal structure of some of the alums.* Phil. Mag. (7) **4** (1927) 688–698.

[7] A. Hargreaves. *Crystal structure of zinc p-toluene sulphonate.* Nature **158** (1946) 620–621.

[8] A. Hargreaves. *The application of the isomorphous replacement method in the determination of centrosymmetric structures.* Acta Crystallogr. **10** (1957) 196–199.

[9] D. Harker. *The determination of the phases of the structure factors of noncentrosymmetric crystals by the method of double isomorphous replacement.* Acta Crystallogr. **9** (1956) 1–9.

[10] G. Kartha. *Isomorphous replacement method in noncentrosymmetric structures.* Acta Crystallogr. **14** (1961) 680–686.

[11] C. M. Mitchell. *Phase determination by the two wavelengths methods of Okaya and Pepinsky.* Acta Crystallogr. **10** (1957) 475–476.

[12] R. Pepinsky and Y. Okaya. *Determination of crystal structures by means of anomalously scattered x-rays.* Proc. Nat. Acad. Sci. Wash. **42** (1956) 286–292.

[13] M. F. Perutz. *Isomorphous replacement and phase determination in noncentrosymmetric space groups.* Acta Crystallogr. **9** (1956) 867–873.

[14] G. N. Ramachandran and S. Raman. *Fourier syntheses for the deconvolution of the Patterson function. Part I. General principles.* Acta Crystallogr. **12** (1959) 957–964.

[15] S. Ramaseshan and K. Venkatesan. *The use of anomalous scattering without phase change in crystal structure analysis.* Curr. Sci. (India) **26** (1957) 352–353.

[16] S. Ramaseshan, K. Venkatesan, and N. V. Mani. *The use of anomalous scattering for the determination of crystal structures—KMnO₄.* Proc. Ind. Acad. Sci. **A46** (1957) 95–111.

[17] J. M. Robertson and I. Woodward. *An x-ray study of the phthalocyanines. Part III. Quantitative structure determination of nickel phthalocyanine.* J. Chem. Soc. (1937) 219–230.

10

Isomorphous replacement: special Fourier syntheses

The alpha and beta syntheses applied to isomorphous crystals

In this chapter we shall be concerned with the extension of the theory of the α and β classes of syntheses to the case of isomorphous crystals. We shall show that it is possible to formulate syntheses with coefficients similar to the general α and β syntheses, using data from isomorphous crystals. These syntheses have the property of revealing the unknown atoms, without having to calculate all the phase angles, as in the conventional methods. We follow here essentially the treatment developed in the laboratory of the authors.[32-34,36,37]

We shall first consider a single pair of isomorphous crystals of the replacement type. Using the notation of Ch. 9, let the structure factors of the two crystals be denoted by F^I and F^{II} and the contributions to these from the replaceable atoms by F_P^I and F_P^{II}. Let the invariant part of the structure be the Q atoms, whose structure factor is F_Q. We assume strict isomorphism, so that we may write

$$F_P^I = |F_P^I| \exp(i\phi_P), \tag{1a}$$

$$F_P^{II} = |F_P^{II}| \exp(i\phi_P), \tag{1b}$$

where the phase factor $\exp(i\phi_P)$ is the same for both. Consider now the quantity

$$D = \delta|F_N|^2 - \delta|F_P|^2 \tag{2}$$

$$= (|F_N^{II}|^2 - |F_N^I|^2) - (|F_P^{II}|^2 - |F_P^I|^2). \tag{3}$$

131

We have the relations

$$F_N^I = F_P^I + F_Q, \qquad F_N^{II} = F_P^{II} + F_Q, \qquad (4)$$

so that

$$|F_N^I|^2 = |F_P^I|^2 + |F_Q|^2 + F_P^I \tilde{F}_Q + \tilde{F}_P^I F_Q, \qquad (5a)$$

$$|F_N^{II}|^2 = |F_P^{II}|^2 + |F_Q|^2 + F_P^{II} \tilde{F}_Q + \tilde{F}_P^{II} F_Q. \qquad (5b)$$

Thus

$$\delta |F_N|^2 = \delta |F_P|^2 + \delta F_P \tilde{F}_Q + \delta \tilde{F}_P F_Q, \qquad (6)$$

hence

$$D = \delta F_P \tilde{F}_Q + \delta \tilde{F}_P F_Q. \qquad (7)$$

We now define the alpha synthesis for isomorphous crystals by

$$\alpha_{is} = D(\delta F_P) \qquad (8)$$

$$= (\delta |F_N|^2 - \delta |F_P|^2)\delta F_P. \qquad (9)$$

Substituting for D from (7), (8) becomes

$$\alpha_{is} = (\delta F_P)^2 \tilde{F}_Q + |\delta F_P|^2 F_Q. \qquad (10)$$
$$\qquad\qquad 3 \qquad\qquad 4$$

This may be compared with the terms of the α_{gen} synthesis, (5) of Ch. 5. Instead of $F_P^2 \tilde{F}_Q$ and $|F_P|^2 F_Q$ designated as 3 and 4, respectively, in (5) of Ch. 5, we now have $(\delta F_P)^2 \tilde{F}_Q$ and $|\delta F_P|^2 F_Q$ in (10) above. Moreover, no terms of the type 1 and 2 of (5) of Ch. 5 occur here. The occurrence of $(\delta F_P)^2$ or $|\delta F_P|^2$ in (10) instead of F_P^2 and $|F_P|^2$ of the α synthesis is only equivalent to a scaling of the coefficients. Thus it is obvious that if we multiply the α_{is} coefficients by

$$K = \frac{f_P}{f_P^{II} - f_P^I}, \qquad (11a)$$

where

$$f_P = \frac{f_P^I + f_P^{II}}{2}, \qquad (11b)$$

we get essentially terms of the type $F_P^2 \tilde{F}_Q + |F_P|^2 F_Q$ where F_P now corresponds to $\sum_j f_P \exp(2\pi i \mathbf{H} \cdot \mathbf{r}_{Pj})$. Thus we may neglect the constant K, which is only a scale factor, and take (10) to represent the α_{is} synthesis.

Thus, it is seen that the terms 3 and 4 in the α_{is} synthesis give rise to peaks in the same positions as 3 and 4 of the α_{gen} synthesis. The elimination of the peaks of the type 1 and 2 is significant, and is obviously due to the fact that more data have been used in the α_{is} synthesis than in the α_{gen} synthesis. The required peaks *4.1* thus occur with a background contributed only by *3.1*, *3.2*, and *4.2* (see Table 1 of Ch. 5). We may also note that the known replaceable-atom peaks are also suppressed. This is obviously due to the fact that the term $|\delta F_P|^2$ is subtracted out of $|\delta F_N|^2$ in D.

The beta synthesis for isomorphous crystals, β_{is}, can be defined in a similar manner. Thus, dividing (7) by $\delta \tilde{F}_P$, we obtain

$$\beta_{is} = \frac{D}{\delta \tilde{F}_P} = \frac{\delta |F_N|^2 - \delta |F_P|^2}{\delta \tilde{F}_P}, \tag{12}$$

which, from (7), becomes

$$\beta_{is} = \underset{7}{F_Q} + \underset{8}{\tilde{F}_Q \exp(2i\phi_P)}. \tag{13}$$

A comparison with (7b) of Ch. 6 shows that the terms *5* and *6* of the β_{gen} synthesis considered there are absent in the β_{is} synthesis, which gives just the required structure F_Q with only a background arising from *8*, which is the convolution of the inverse structure of the Q atoms with the phase-squared structure of the P atoms. We may write down the coefficient of the β_{is} synthesis more explicitly in view of its importance:

$$\beta_{is} = \frac{(|F_N^{II}|^2 - |F_N^{I}|^2) - (|F_P^{II}|^2 - |F_P^{I}|^2)}{|F_P^{II} - F_P^{I}|} \exp(i\phi_P). \tag{14}$$

The quantities used in the above synthesis are all known. The first term in the numerator is the difference in intensity between the two crystals. The second term is the difference in the squares of the structure amplitude of the replaceable atoms in the two crystals, which can be calculated. So also, the term in the denominator and $\exp(i\phi_P)$ are readily calculated, since the positions of the P atoms are known.

It is interesting to note that the right-hand side of (13) is identical to (8) of Ch. 9, which means that the β_{is} synthesis is just the double-phased Fourier synthesis, the two ambiguous values of the phase used here corresponding to that of the Q atoms. When, however, the P group is centrosymmetric, $\phi_P = 0$ or π, and $\exp(2i\phi_P) \equiv 1$, so that the β_{is} synthesis becomes

$$\beta_{is} = F_Q + \tilde{F}_Q \tag{15}$$

Thus, when the structure as a whole is noncentrosymmetric, but the P atoms alone are centrosymmetric, the β_{is} synthesis yields the structure F_Q duplicated by its inverse structure $\tilde{F}_Q$, the two sets of images having a center of symmetry at the point of inversion of the P group. This result is also obvious from our earlier analysis in Ch. 9. When the structure as a whole is centrosymmetric, $F_Q \equiv \tilde{F}_Q$ and hence we see from (15) that the β_{is} synthesis gives the structure

of the Q atoms completely, with no extra background. This is also under-
standable, since we know that, for a centrosymmetric crystal, the sign
determination is unique. However, what is of particular interest is that the
β_{is} synthesis can be performed as a simple Fourier synthesis with the coefficients defined by (14), without having to calculate the phases or signs and
trying to pick out the correct one.

The extension of the β_{is} synthesis to the case of more than one pair of
crystals can also be made readily. For convenience, we shall consider isomorphism of the addition type considered in the last chapter. Let F^I, F^{II}, and
F^{III} be the coefficients of the three crystals, with $f_P^I = 0$, and f_P^{II}, and $f_{P'}^{III}$
being the scattering powers of the heavy atoms, in the derivatives II and III.
We have used the symbol P' in $f_{P'}^{III}$ to denote that the P' atoms do not occupy
the same positions as the P atoms in crystal II. We can now obtain two β_{is}
syntheses:

$$\beta_{is}^{I, II} = \frac{(|F^{II}|^2 - |F^I|^2) - |F_P^{II}|^2}{|F_P^{II}|} \exp(i\phi_P), \tag{16}$$

and

$$\beta_{is}^{I, III} = \frac{(|F^{III}|^2 - |F^I|^2) - |F_{P'}^{III}|^2}{|F_{P'}^{III}|} \exp(i\phi_{P'}). \tag{17}$$

These two syntheses will respectively correspond to

$$F_Q + \tilde{F}_Q \exp(2i\phi_P), \tag{18a}$$

and

$$F_Q + \tilde{F}_Q \exp(2i\phi_{P'}). \tag{18b}$$

Thus if we perform a synthesis $[\beta_{is}^{I, II} + \beta_{is}^{I, III}]$ as coefficients, it will lead to the
structure

$$2F_Q + [\tilde{F}_Q \exp(2i\phi_P) + \tilde{F}_Q \exp(2i\phi_{P'})]. \tag{19}$$

It is to be noted that it is necessary that the P and P' atoms should occupy
different positions in the two crystals relative to the Q atoms in crystal I,
for otherwise the two terms within brackets in (19) become identical, which
is equivalent to performing a β_{is} synthesis with a single pair and merely
multiplying the synthesis (18a) by a factor of two. The combined synthesis
(19), however, makes the required peaks in the Q structure twice as strong,
while the background peaks are not increased in strength, as they are distributed. The latter is because the individual small peaks in the two structure
$\tilde{F}_Q \exp(2i\phi_P)$ and $\tilde{F}_Q \exp(2i\phi_{P'})$ will not be coincident except by accident.
This could also be seen in another way. The combination synthesis (19) is

equivalent to performing the double-phased syntheses for the two pairs, and adding the two. Then if the phase angle ambiguity is taken to be

$$\phi_Q = \phi_P \pm \theta, \tag{20a}$$

and

$$\phi_Q = \phi_{P'} \pm \theta', \tag{20b}$$

one of the two solutions in each of (20a) and (20b) will coincide. Thus the combination β_{is} synthesis (19) is equivalent to performing a Fourier synthesis with all the *three* phase angles, but with the correct phase angle having twice the weight compared to the two incorrect ones. This synthesis, however, is not equivalent to one using uniquely the correct phase angle, which is the common solution for the two pairs of crystals.

Tests of the isomorphous syntheses

In view of the equivalence of the β_{is} synthesis with the double-phased synthesis, which has been commonly used with isomorphous crystals, examples of successful application of the latter may be taken essentially to be tests for the validity of the β_{is} synthesis. Thus the double-phased synthesis was applied by Bokhoven, Schoone, and Bijvoet[5] in the solution of the structure of strychnine. In this case, the replaceable atoms formed a centrosymmetric configuration, while the structure as a whole was noncentrosymmetric, so that there was a duplication of the structure by inversion in the resultant synthesis. Both double-phased syntheses, (7) and (8) of Ch. 9, were tested on a hypothetical noncentrosymmetric case by Kartha.[18] The replaceable atoms formed a noncentrosymmetric constellation and the structure was revealed without any duplication, except for a background arising out of the second term in the right-hand side of (7) and (8) of Ch. 9. In the above two examples, however, the syntheses were performed by actually calculating the phase angles and feeding in the two ambiguous values. Synthesis (8) of Ch. 9 in this case would correspond to suppressing the replaceable atoms by subtracting the known structure factors of the P group from the double-phased Fourier in which $|F_N|$ is used.

A Fourier synthesis making use of both phases has also been used by Blow and Rossmann[3,42] for the complex protein structures of myoglobin and hemoglobin. This single isomorphous method led to an interpretable Fourier map, when the replaceable group formed an acentric configuration. However, Blow and Rossmann used a weighting scheme for the two phases, based on the probability distribution of the phase angles. We shall consider the aspects of weighting of the phases determined by isomorphous replacement in detail in a later section in this chapter.

The method of using the β_{is} synthesis directly by calculating the coefficients, without determining the phase angles, was first tested by Ramachandran and Ayyar.[33] They tried it on a semihypothetical model by taking the actual structure of P_4S_5 and replacing three of the sulfur atoms by atoms of three times their scattering power. The β_{is} synthesis was calculated for the non-centrosymmetric c-axis projection for this hypothetical isomorphous pair and this is shown in Fig. 1. It may be seen that all the six remaining atoms

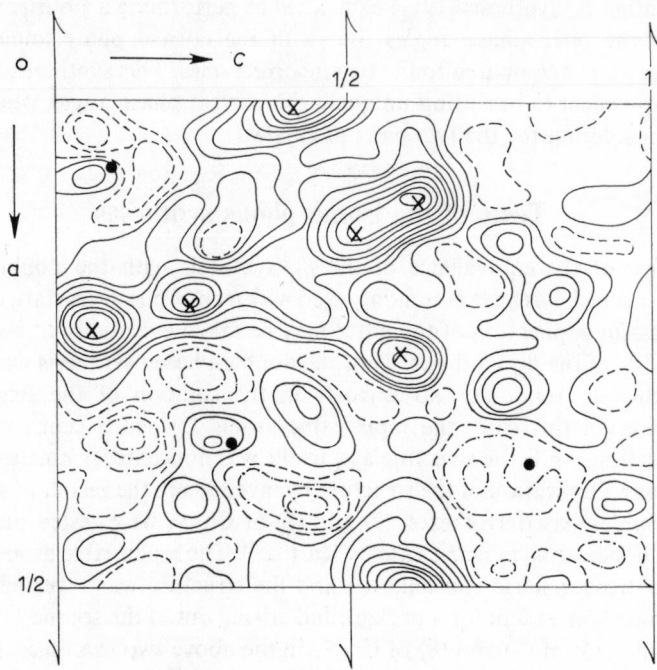

Fig. 1. The beta-isomorphous synthesis for P_4S_5 structure with three replaceable atoms (marked as dots). Note that the peaks at the replaceable atoms are suppressed and the rest of the structure (crosses) has come out strongly.

have come up quite clearly. Also the replaceable atoms have been suppressed, as is to be expected from theory.

An example of the test reaching more closely to an actual practical situation is to be found in the case of the pair, arginine hydrochloride monohydrate and arginine hydrobromide monohydrate.[25] The β_{is} synthesis was tested on a projection (Fig. 2) using the actual intensity data from the pair of crystals. It will be seen that the rest of the structure has come out clearly in Fig. 2. There are two molecules of $C_6H_{14}O_2N_4,HX,H_2O$ in the asymmetric unit,

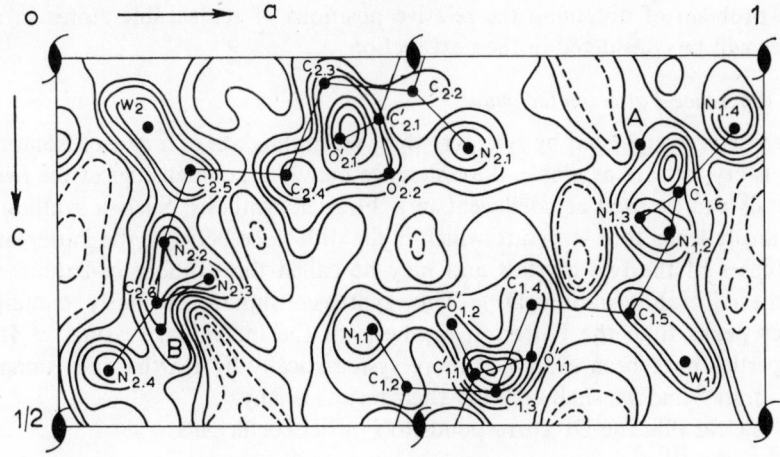

Fig. 2. The beta-isomorphous synthesis for the pair L-arginine HBr · H$_2$O and L-arginine HCl · H$_2$O. Heavy-atom positions are marked *A* and *B*. There are two molecules in the asymmetric unit; atoms are designated by subscripts *n*, *j*, where $n = 1$ or 2, denotes the molecule and *j* the serial number of atom of each type, e.g., C$_{2.6}$ or O$_{1.1}$.

the space group being P2$_1$, so that the replaceable atoms (X = Cl,Br) formed a noncentrosymmetric configuration. These two structures had been solved earlier by using the simple β_{gen} synthesis on the hydrobromide compound.[26] This example shows how the β_{is} synthesis could be expected to work in an actual case, where the intensity data of two compounds may be subject to possible errors and the two unit cells may not be absolutely identical.

Determination of heavy-atom positions

The method of isomorphous replacement has invariably been associated with the heavy-atom method, for the atoms involved in the replacement are usually heavy atoms. The problem of the determination of the replaceable-atom positions is not difficult in such cases, since it could be done by using the simple Patterson synthesis, and picking out from it the strong heavy-atom interactions. However, the problem assumes considerable importance in complex structures, such as in the case of protein crystals and their derivatives, in which the heavy-atom interactions do not stand out clearly in the simple Patterson map. Another related problem also exists in such cases, namely that of obtaining the relative positions of the heavy atoms in a series of multiple isomorphous crystals. In this section we shall consider methods of finding the positions of the replaceable atoms in a given pair of crystals.

The problem of obtaining the relative positions of replaceable atoms in a series will be considered in the next section.

The difference-Patterson technique

It was first suggested by Buerger[7] that, when the data of a pair of isomorphous crystals are available, the difference between the intensities of the two crystals can be used as coefficient in a Fourier synthesis. Such a synthesis obviously leads to a structure which is the difference between the Patterson syntheses of the two crystals and may be called the "difference-Patterson synthesis" (abbreviation DP). This synthesis should obviously contain fewer peaks than the Patterson synthesis of the individual crystals.[7,8] Its properties have been examined more systematically by Kartha and Ramachandran[23] and we shall consider their results briefly.

It is clear that the DP corresponds to Fourier coefficients

$$(|F^{II}|^2 - |F^{I}|^2) = (F_P^{II} + F_Q)(\tilde{F}_P^{II} + \tilde{F}_Q) - (F_P^{I} + F_Q)(\tilde{F}_P^{I} + \tilde{F}_Q), \qquad (21)$$

where the notation follows that introduced in Ch. 9. On simplifying the right-hand side of (21) we get

$$|F^{II}|^2 - |F^{I}|^2 = \delta|F_P|^2 + \delta F_P \tilde{F}_Q + \delta \tilde{F}_P F_Q, \qquad (22)$$
$$\phantom{|F^{II}|^2 - |F^{I}|^2 = } a \qquad\quad b \qquad\quad c$$

where δ stands for the difference of the corresponding quantities in the two crystals. The first term a on the right-hand side of (22) represents the difference of the Pattersons of the P group of atoms in the two crystals. Since the positions of the peaks are identical in both crystals, it will contain peaks corresponding only to the vectors between the P atoms, but their strengths will be $(f_P^{II})^2 - (f_P^{I})^2$, assuming all the P atoms to be alike. The terms b and c would give peaks corresponding to the interactions between the P atoms and the Q atoms. The strengths of these will be $(f_{Pj}^{II} - f_{Pj}^{I})f_{Qk}$. It may be noticed that no interactions of the Q-Q type are present in the DP. Considered from the geometrical point of view, the DP contains $2PQ$ peaks as compared with N^2 for the individual Pattersons. These $2PQ$ peaks consist of the P images of the Q atoms as seen from the P atoms (PQ in number) and an additional equal number obtained by duplication of these by inversion about the origin.

There is a relative enhancement of the strengths of the P–P type of peaks in the DP relative to the background, as compared with the ordinary Patterson. In the former, the ratio of a P–P to a P–Q type of interaction is

$$\delta(f_P^2)/f_Q \, \delta f_P = (f_P^{I} + f_P^{II})/f_Q$$

(in the equal P atom case) while in the ordinary Patterson, this ratio is just f_P/f_Q. The former is always greater than the latter, so that this property may facilitate identification of the replaceable atom vectors by computing the DP synthesis.

The difference Patterson was first applied by Frueh[13] in an interesting application to order–disorder transformation problems and in the crystal-structure determination of tourmaline.[12] It was also tried by Sasisekharan[45] on a pair of crystal structures whose structure had been established by a different method. An attempt to use it on the pair caffeine and theophylline by Sutor[55] led to an uninterpretable map. This particular pair is an example of isomorphism of the addition type, in which one of the compounds differs from the other by a methyl group. Since the molecule was small, the addition of a methyl group led to appreciable changes in the unit-cell dimensions and in the positions of the common atoms. This was probably the reason why the method failed in this case.

More generally, the locations and strengths of the peaks in the DP, when the atoms are unlike, are summarized in Table 1.

Table 1   Peaks occurring in the difference Patterson

Type	Vector notation	Location	Strength
$P\text{–}P$	$[P_i P_j]$	$\mathbf{r}_{Pj} - \mathbf{r}_{Pi}$	$f_{Pi}^{II} f_{Pj}^{II} - f_{Pi}^{I} f_{Pj}^{I}$
$P\text{–}Q$	$[P_j Q_k]$	$\mathbf{r}_{Qk} - \mathbf{r}_{Pj}$	$f_{Qk}[f_{Pj}^{II} - f_{Pj}^{I}]$
$P\text{–}Q$	$[Q_k P_j]$	$\mathbf{r}_{Pj} - \mathbf{r}_{Qk}$	$f_{Qk}[f_{Pj}^{II} - f_{Pj}^{I}]$
$Q\text{–}Q$	$[Q_k Q_l]$	Absent	

Location of heavy-atom vectors on the difference Patterson by the principle of maximum superposition

The strength of the peaks in the DP diagram corresponding to the termini of the P–P vectors may not be large if the scattering factors of the P atoms are of the same order of magnitude as those of the Q atoms. In fact, this is precisely the situation for which an effective method is required, since otherwise even the ordinary Patterson synthesis would be sufficient. In particular, it may be noticed that, if the isomorphism is of the addition type, there is no advantage gained at all with the DP since f_{Pj}^{I} are all zero, and both the DP and the ordinary Patterson have the ratio f_{P}^{II}/f_{Q} for the peak heights of a P–P and a P–Q type of interaction (except for an appreciable reduction of the background in the DP). The method also progressively decreases in its power as the number of Q atoms increases, since the sum of the images of the Q atoms tends to swamp the peaks corresponding to the P–P vectors. This is the typical type of difficulty met with in protein crystal analysis where, usually, an isomorphous series of the addition type is available. A method known as the principle of maximum superposition[31,44] applicable to DP, is likely to prove useful in such a situation. An essentially similar method has been suggested by Raman and Lipscomb.[38]

The method is best understood through the theory of images. As may be seen from Table 1, the peaks in the DP may be characterized as follows:

$$\text{strength } \delta f_P(f_{Pj}^{\text{I}} + f_{Pi}^{\text{II}}) \quad \text{at} \quad \mathbf{r}_{Pi}^{\text{II}} - \mathbf{r}_{Pj}^{\text{I}}, \tag{23a}$$

$$\text{strength } \delta f_{Pi} f_{Qj} \quad \text{at} \quad \mathbf{r}_{Pi} - \mathbf{r}_{Qj}, \tag{23b}$$

$$\text{strength } f_{Qi} \delta f_{Pj} \quad \text{at} \quad \mathbf{r}_{Qi} - \mathbf{r}_{Pj}, \tag{23c}$$

where $\delta f_P = f_P^{\text{II}} - f_P^{\text{I}}$, and all the P atoms in each crystal have been assumed to be alike. If we now shift the origin of the DP by a vector $[P_1 P_2]$, then we have the following coincidences between the shifted and unshifted diagrams:

$$\{P_j P_1\} + [P_1 P_2] = \{P_j P_2\}, \tag{24a}$$

$$\{P_2 P_j\} + [P_1 P_2] = \{P_1 P_j\}, \tag{24b}$$

$$\{Q_j P_1\} + [P_1 P_2] = \{Q_j P_2\}, \tag{24c}$$

$$\{P_2 Q_j\} + [P_1 P_2] = \{P_1 Q_j\}. \tag{24d}$$

The number of peaks in the first two sets are P each, while they are Q each in the last two sets. On the other hand, if we shift the origin by a vector of the type $[PQ]$, say $[P_1 Q_1]$, we obtain only the following superpositions:

$$\{P_j P_1\} + [P_1 Q_1] = \{P_j Q_1\}, \tag{25a}$$

$$\{Q_1 P_j\} + [P_1 Q_1] = \{P_1 P_j\}, \tag{25b}$$

in each of which the number of superpositions is P. In other words, because of the fact that the Q–Q type of vectors are absent from the DP, the shifts corresponding to a P–Q vector produce only partial images where the peaks superpose, while a shift of the type P–P produces two complete images—one as seen from P_1 and the other the inverse of the whole structure as seen from P_2. It is obvious that the ratio of the number of coincidences for a P–P and a P–Q vector shift is $(P + Q)/P$. When P is small and Q is large, this ratio becomes large. This is precisely the situation we are concerned with in the determination of heavy-atom vectors in complex structures such as the proteins. This property can therefore be made use of effectively, to distinguish between a $[PP]$ and a $[PQ]$ vector.

We can use now the cumulative image-seeking function discussed in Ch. 3. The method of applying this in practice is as follows. The DP is calculated in the usual way over a grid and the vector shift is made to the various grid points which are the likely sites of P–P interactions. For each vector shift, one of the image-seeking functions (e.g., product or minimum function) is computed. The integral of this function over the entire unit cell is calculated and this is entered on a map at a point corresponding to the vector shift employed. If necessary, the entire unit cell may be covered in this way and we obtain a map of the cumulative superposition function of the DP. This map will obviously contain large maxima corresponding to a P-P vector, while the maxima corresponding to a P-Q vector will have considerably lower strength. The contrast will be larger the more the number of Q atoms.

As mentioned above, we may use the minimum or the product function for obtaining the cumulative function, preferably the former. The sum function cannot be used, since the integral for every vector is a constant and equal to twice the integral of the value of the DP function over the unit cell. The above is termed the "principle of maximum superposition."[44] Essentially the same procedure has been suggested by Raman and Lipscomb.[38]

This method of approach will be practical for complicated structures only with the availability of very large computers, since it involves an enormous amount of calculation. A test in a simple case is given in a later section of this chapter.

The modulus difference squared $(MD)^2$ synthesis

In the previous subsections we were concerned with the synthesis which uses the difference between the intensities of a pair of isomorphous crystals. We shall now consider a synthesis which uses as coefficients the square of the difference in the moduli, viz., the quantity $(|F_N^{II}| - |F_N^{I}|)^2$. We may call this the modulus-difference squared synthesis (abbreviation, MD-squared synthesis). In this connection, we may also introduce another terminology originally due to Freuh,[13] namely the Patterson-difference synthesis (PD for short), to denote the synthesis using the coefficients $|F_N^{II} - F_N^{I}|^2$. The name is obvious, since it corresponds to the Patterson of the difference structure. The PD is useful for purposes of discussion, since it corresponds theoretically to the ideal case of the Patterson of the difference structure. It cannot, however, be calculated in a case in which the structure is unknown, since it requires the knowledge of the phases of the two crystals.

In our present discussion we would like to examine how closely the MD-squared synthesis would correspond to the PD. The difference structure $(F_N^{II} - F_N^{I})$, $\equiv \delta F_N$, would contain peaks corresponding to the replaceable atoms, and their strengths will be obviously δf_P. Also, if the isomorphism is ideal, $\delta F_N = \delta F_P$, so that $|\delta F_N|^2$ is equal to $|\delta F_P|^2$ for the pair of crystals, or the PD structure is identical with the Patterson of the P atoms, but with peak strengths decided by the difference structure. Thus the PD synthesis will have peaks of

$$\text{strength} (\delta f_{Pi})(\delta f_{Pj}) \quad \text{at} \quad (\mathbf{r}_{Pi} - \mathbf{r}_{Pj}). \tag{26}$$

We shall first consider the centrosymmetric case. If the structure is complex (i.e., the Q atoms are large in number), the contributions from the P atoms to the total intensity in the two crystals will be small compared to that from the rest of the structure, and, therefore, both F_N^{II} and F_N^I are likely to have the same sign. Hence to a fairly good degree of approximation, we may write

$$|F_N^{II} - F_N^I| = \|F_N^{II}| - |F_N^I\| . \tag{27}$$

Under these conditions, the MD-squared synthesis is a sufficiently good approximation to the PD. Such a synthesis, although not exactly in this form, was first applied by Frueh[13] in an order–disorder transformation problem.

We might expect that this property of approximate equality of the MD-squared and the PD syntheses would also hold good for the noncentro-symmetric case.[1] Rossmann[41] showed (although using a rather oversimplified picture) that, even in the noncentrosymmetric case, this synthesis should contain peaks corresponding to the P–P vectors and thus give an approximation to the PD. This may be seen as follows. Referring to Fig. 3, the vector δF_P makes, in general, an angle θ^I with the vector F_N^I, as a result of which it is different from $|F_N^{II}| - |F_N^I|$. It is clear that, when the observed difference is large, it is likely that F_N^I, F_N^{II}, and δF_P are nearly collinear, so that $|\delta F_N| \simeq |\delta F_P|$. The two will differ appreciably only when $|\delta F_N|^2$ is small. The MD-squared synthesis is thus equivalent to giving lower weights to such reflections, which is reasonable, since the corresponding coefficients tend to be small in the Fourier synthesis. We could therefore expect that it should, as a first approximation, tend to the PD synthesis. It may also be noted that these conditions are more favorable the smaller the contribution from the P group, compared with the rest of the structure.

We may make the argument more quantitative as follows. Referring to Fig. 3, and assuming $\delta F_P \ll F_N^I$ or F_N^{II}, we have the relation (see also (17) of Ch. 13)

$$|\delta |F_N\| \simeq |\delta F_P| \cos \theta^I, \tag{28}$$

where $\theta^I = (\phi_N^I - \phi_P)$. Thus

$$|\delta |F_N\|^2 \simeq |\delta F_P|^2 \cos^2 \theta^I$$

$$= \tfrac{1}{2} |\delta F_P|^2 + \tfrac{1}{2} |\delta F_P|^2 \cos 2\theta^I \tag{29a}$$

$$= \tfrac{1}{2} |\delta F_P|^2 + \tfrac{1}{4} |\delta F_P|^2 \{\exp{(2i\phi_P)} \exp{(-2i\phi_N^I)}$$

$$+ \exp{(-2i\phi_P)} \exp{(2i\phi_N^I)}\}. \tag{29b}$$

It follows from (29b) that, as a first approximation, the MD-squared synthesis should contain peaks corresponding to the PD, but with half the strength, arising out of the first term in (29b). The remaining terms in (29b) represent the

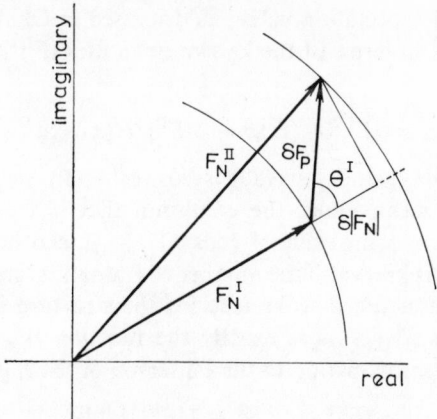

Fig. 3. Relation between $|\delta F_N|$ and $|\delta F_P|$ for a
pair of isomorphous crystals.

convolution of the Patterson with the phase-squared structures of the P atoms
and the N atoms and their inverses. These can be expected to be widely
dispersed and will thus contribute only to a general background, against
which the concentrated peaks of the PD should stand out.

A modified form of the MD-squared synthesis based on statistical considerations has also been suggested.[17,46] Considering the isomorphism of the
addition type (Fig. 4), the ideal Patterson for revealing the P atoms is obviously

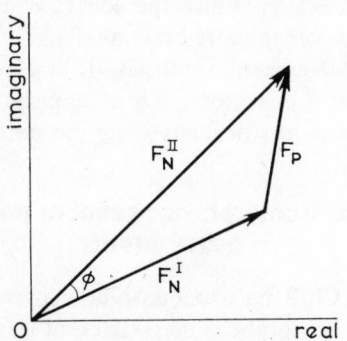

Fig. 4. Relation between vectors for
a pair of (addition-type) isomorphous crystals.

$|F_P|^2$. In the absence of precise values of these coefficients, it would be best
to use values which are the nearest approximations to these quantities. We
may therefore apply for this purpose the principle of replacing the unknown

by its conditional expectation value, as discussed in Ch. 8. From Fig. 4, we may express $|F_P|^2$ in terms of the known quantities $|F^I|^2$, $= |F_Q|^2$, and $|F^{II}|^2$ in the form

$$|F_P|^2 = |F^I|^2 + |F^{II}|^2 - 2|F^{II}||F^I|\cos(\phi^{II} - \phi^I) \qquad (30)$$

The only unknown quantity in (30) is $\cos(\phi^{II} - \phi^I)$ and therefore we may replace it by its mean under the condition that $|F^I|$ and $|F^{II}|$ are given. Calling $(\phi^{II} - \phi^I) = \phi$, the value of $\langle\cos\phi\rangle_{|F^{II}|, |F^I|}$ can be worked out if the distribution of F_P is known. If the number of P atoms is large, so that Wilson's statistics could be assumed to be true for the structure factors F_P, it turns out[17,46] that $\langle\cos\phi\rangle_{|F^{II}|, |F^I|}$ is exactly the function W_A obtained earlier in Ch. 8. The best approximation to the Patterson of the P group is thus

$$|F^I|^2 + |F^{II}|^2 - 2|F^{II}||F^I|W_A. \qquad (31)$$

When W_A tends to 1, the coefficient tends to that of the MD-squared synthesis while, for intermediate values, it differs from $(|F^{II}| - |F^I|)^2$. Expression (31) may therefore be considered as a modified, or weighted, MD-squared synthesis. A test of this synthesis is described in a later section of this chapter.

We have thus shown that the square of the difference in moduli of two isomorphous crystals, or a suitably weighted modification of this, leads to the Patterson of the difference structure as a first approximation. It would appear, from our earlier analysis of the properties of functions of the form $|F|^n$, that we might also use a function which is the square of the difference in intensities and hope to locate the P–P vectors. Such a synthesis using $[|F^{II}|^2 - |F^I|^2]^2$ as coefficients, which is also the square of the difference Patterson (DP-squared), is amenable to more exact analysis[38] from a theoretical point of view than the MD-squared synthesis. It does contain positive peaks corresponding to the P–P vectors, but it appears that the MD-squared synthesis works better in practical cases, e.g., in protein crystallography.

Multiple isomorphism: relative positions of heavy atoms

It was mentioned in Ch. 9 that a unique solution of the phase problem exists[15] if we have a series of isomorphous derivatives, at least three in number. One of the conditions for such a unique solution to be possible is that the replaceable-atom structure factors of two different crystals should not be collinear. This necessarily means that, relative to, say, crystal I as reference, the replaceable atoms should occupy different locations in crystals II and III. The problem of finding out the relative positions of the P atoms in crystals II and III is an important one. In crystals where there are intersecting symmetry axes, this is not particularly difficult. However, in cases where there is a degree of freedom for the choice of origin in individual crystals (as in space group $P2_1$, in which the origin can be chosen anywhere along the 2_1 axis), the problem exists of finding the relative positions of the replaceable atoms with reference to a chosen origin (say, as fixed in crystal I). In order to simplify the discussion, we shall consider throughout this section the case of a series of three isomorphous crystals of the *addition* type—a case of particular interest in protein crystallography. To be specific, the atoms additional to the first crystal, I, are P^{II}, in crystal II, and P^{III}, in crystal III, and the three structure factors are denoted by F^I, F^{II}, F^{III}. The relations between F^I, F^{II}, and F^{III} are given by (15a), (15b), and (15c) of Ch. 9 and are shown diagrammatically in Fig. 5.

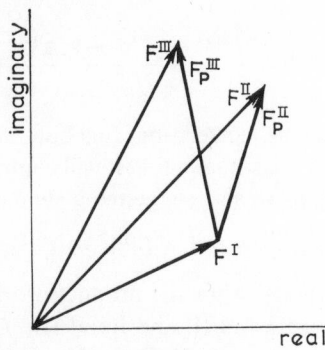

Fig. 5. Relation between vectors in (addition-type) double isomorphous crystals.

Harker[15] suggested that the correlation between the set of atoms P^{II} and P^{III} could be done by making use of the phase-circle diagram shown in Fig. 2 of Ch. 9. Given the magnitudes of $|F^I|$, $|F^{II}|$, $|F^{III}|$, $|F_P^{II}|$, and $|F_P^{III}|$, the angle between the vectors $-F_P^{II}$ and $-F_P^{III}$ in that figure could be obtained by trial and error, so that all the three circles intersect at a point. If this is done for a few selected reflections, then the required correlation between the origins of the unit cells in crystals II and III can be made.

Correlation functions

Bragg[6] suggested a simple graphical method of estimating the relative positions of heavy atoms, which makes use of the observed differences $(|F^{III}| - |F^I|)$ and $(|F^{II}| - |F^I|)$. This was used in the case of myoglobin derivatives. The method loses its simplicity when the number of substituents increases, as well as when the symmetry in the unit cell increases the number of equivalent positions. More effective methods involving correlation syntheses have been suggested later.

Perutz[29] proposed two functions for the location of the relative positions of the heavy atoms in two derivatives. He considered a simplified case of having one heavy atom attached to the parent crystal (crystal I) in space group P1, to form one derivative (crystal II), and similarly another single heavy atom attached at some other site to form a second derivative (crystal III). Since the P group consists of just one atom in each case, we shall represent their structure factors by f_P^{II} and f_P^{III}, respectively. The relations between F^I, F^{II}, and F^{III} are given by (15a), (15b), and (15c) of Ch. 9.

Defining two quantities

$$\mathscr{B}^{II} = \frac{|F^{II}|^2 - |F^I|^2 - (f_P^{II})^2}{2f_P^{II}}, \tag{32a}$$

$$\mathscr{B}^{III} = \frac{|F^{III}|^2 - |F^I|^2 - (f_P^{III})^2}{2f_P^{III}}, \tag{32b}$$

Perutz suggested[29] that the correlation function $\mathscr{B}^{II}\mathscr{B}^{III}$ can be used as coefficients in a Fourier synthesis, and that it will contain peaks corresponding to the vectors $\pm(\mathbf{r}_P^{II} - \mathbf{r}_P^{III})$. He also suggested a synthesis using

$$(|F^{II}|^2 - |F^I|^2)(|F^{III}|^2 - |F^I|^2)$$

as coefficients. This synthesis is just the product of the difference Pattersons of the two pairs of crystals I and II, and I and III. Two other functions derived from the above were added to the list by Hoppe[16] and these correspond, respectively, to using the products $(|F^{II}|^2 - |F^I|^2 - f_P^{II2})(|F^{III}|^2 - |F^I|^2 - f_P^{III2})$ and $(\mathscr{B}^{II}/|F^I|)(\mathscr{B}^{III}/|F^I|)$.

It is possible to consider the theory of all the above syntheses without making any simplifying assumption with regard to the number of heavy atoms P^{II} and P^{III} in the unit cell. Thus, consider the following four functions $\mathscr{A}^{II}\mathscr{A}^{III}$, $\mathscr{B}^{II}\mathscr{B}^{III}$, $\mathscr{C}^{II}\mathscr{C}^{III}$, $\mathscr{D}^{II}\mathscr{D}^{III}$, where

$$\mathscr{A}^{II} = |F^{II}|^2 - |F^I|^2 - |F_P^{II}|^2, \qquad \mathscr{A}^{III} = |F^{III}|^2 - |F^{II}|^2 - |F_P^{III}|^2, \tag{33a}$$

$$\mathscr{B}^{II} = \frac{\mathscr{A}^{II}}{2|F_P^{II}|}, \qquad \mathscr{B}^{III} = \frac{\mathscr{A}^{III}}{2|F_P^{III}|}, \tag{33b}$$

$$\mathscr{C}^{II} = \frac{\mathscr{A}^{II}}{2|F^I||F_P^{II}|}, \qquad \mathscr{C}^{III} = \frac{\mathscr{A}^{III}}{2|F^I||F_P^{III}|}, \tag{33c}$$

$$\mathscr{D}^{II} = |F^{II}|^2 - |F^I|^2, \qquad \mathscr{D}^{III} = |F^{III}|^2 - |F^I|^2. \tag{33d}$$

We have used a notation different from that used by Perutz[29] and Hoppe,[16] so as to agree with the general notation used in this monograph. Further, we have assumed that there can be any number of heavy atoms in the unit cell, so that its contribution is given by F_P rather than by f_P. When F_P is replaced by f_P, all these products will reduce to those discussed in the last paragraph for the case of single heavy atoms.

We shall consider in detail the synthesis $\mathscr{A}^{II}\mathscr{A}^{III}$, with $\mathscr{A}^{II}$, $\mathscr{A}^{III}$ defined by (33a), since the interpretation of this is fairly straightforward. Since $F^{II} = F^I + F_P^{II}$, we have

$$\mathscr{A}^{II} = (F^I + F_P^{II})(\tilde{F}^I + \tilde{F}_P^{II}) - |F^I|^2 - |F_P^{II}|^2$$
$$= F_P^{II}\tilde{F}^I + F^I\tilde{F}_P^{II}. \tag{34}$$

The coefficient $\mathscr{A}^{II}\mathscr{A}^{III}$ is thus given by

$$\mathscr{A}^{II}\mathscr{A}^{III} = (F_P^{II}\tilde{F}^I + F^I\tilde{F}_P^{II})(F_P^{III}\tilde{F}^I + F^I\tilde{F}_P^{III})$$
$$= \underset{a}{F_P^{II}F_P^{III}(\tilde{F}^I)^2} + \underset{b}{F_P^{II}\tilde{F}_P^{III}|F^I|^2}$$
$$+ \underset{c}{\tilde{F}_P^{II}F_P^{III}(F^I)^2} + \underset{d}{\tilde{F}_P^{II}\tilde{F}_P^{III}(F^I)^2}. \tag{35}$$

It is readily seen that the structure given by d is the inverse of that given by a, and similarly that given by c and b are inverses. Hence, it is enough to consider a and b. The strongest peak in these originates from term b and is that arising from the convolution of the origin peaks of $|F^I|^2$ with the peaks at $(\mathbf{r}_{Pi}^{II} - \mathbf{r}_{Pj}^{III})$ in the structure $F_P^{II}\tilde{F}_P^{III}$. Since the latter are of strength $f_{Pi}^{II}f_{Pj}^{III}$, the principal peaks of the correlation synthesis $\mathscr{A}^{II}\mathscr{A}^{III}$ are of strength $S_N^{12}f_{Pi}^{II}f_{Pj}^{III}$ at $(\mathbf{r}_{Pi}^{II} - \mathbf{r}_{Pj}^{III})$. Since the three structures which are convolved in the structure given by the term a contain atoms at noncoincident locations, the resultant peaks will be distributed and will contribute only to the background. In particular, it may be noted that $(\tilde{F}^I)^2$ does not have an origin peak, as it is only the square of the inverse structure $\tilde{F}^I$, so that none of the background peaks will have a strength comparable to the strong correlation peaks given by the term b.

It is easy to see that the special case in which the P^{II} and P^{III} groups consist of single atoms would lead to just one pair of interaction peaks at $\pm(\mathbf{r}_P^{II} - \mathbf{r}_P^{III})$, as mentioned earlier. We shall not work out the theory of the other syntheses $\mathscr{B}^{II}\mathscr{B}^{III}$, $\mathscr{C}^{II}\mathscr{C}^{III}$ and $\mathscr{D}^{II}\mathscr{D}^{III}$, but their nature follows from the discussion of the theory of the correlation synthesis $\mathscr{A}^{II}\mathscr{A}^{III}$. It appears that, in general, these syntheses are likely to have more background.

The syntheses $\mathscr{B}^{II}\mathscr{B}^{III}$ and $\mathscr{C}^{II}\mathscr{C}^{III}$ take a particularly simple form in terms of the phase angles. Denoting the phase angles of F^I, F_P^{II}, and F_P^{III} by ϕ^I, ϕ^{II}, and ϕ^{III} respectively, (34) takes the simple form

$$\mathscr{A}^{II} = 2|F_P^{II}||F^I| \cos(\phi_P^{II} - \phi^I). \tag{36a}$$

Similarly,

$$\mathscr{A}^{III} = 2|F_P^{III}||F^I| \cos(\phi_P^{III} - \phi^I). \tag{36b}$$

It follows from this that the coefficient $\mathscr{C}^{II}\mathscr{C}^{III}$ is just equal to $\cos(\phi_P^{II} - \phi^I)$ $\cos(\phi_P^{III} - \phi^I)$. This suggests that another function $\mathscr{G}^{II}\mathscr{G}^{III}$, where

$$\mathscr{G}^{II} = |F_P^{II}|\,\mathscr{C}^{II} = |F_P^{II}|\cos(\phi_P^{II} - \phi^I), \tag{37a}$$

$$\mathscr{G}^{III} = |F\,_P^{III}|\,\mathscr{C}^{III} = |F_P^{III}|\cos(\phi_P^{III} - \phi^I) \tag{37b}$$

should also reveal the correlation vectors $(\mathbf{r}_{Pi}^{II} - \mathbf{r}_{Pj}^{III})$. This function has been used by Blow[1] and by Bodo et al.[4] in the determination of the relative y coordinates of the heavy atoms in the derivatives of hemoglobin and myoglobin, respectively.

The MD-squared synthesis as a correlation synthesis

The correlation functions discussed above all involve terms which are derived from the difference Patterson and related functions. Since they involve the intensities in general, the structures generated by syntheses using the resultant functions are capable of fairly simple interpretations from theory. It is found in practice, however, that in a number of situations the use of functions of the structure amplitudes, rather than the intensities, offers good approximations. For exmaple, we have seen earlier that the MD-squared synthesis is, to a good degree of approximation, equal to the Patterson of the difference structure.

A synthesis similar to this was used by Rossmann[41] as a correlation function. Referring to Fig. 5, suppose we compute the synthesis $(|F^{III}| - |F^{II}|)^2$. We can write this in the form

$$(|F^{III}| - |F^{II}|)^2 = |F^{III}|^2 + |F^{II}|^2 - 2|F^{III}|\,|F^{II}|. \tag{38}$$

The first two terms on the right-hand side of (38) correspond to the Patterson syntheses of the crystals II and III, respectively. They will therefore contain positive peaks corresponding to the heavy-atom vectors in the two crystals, considered separately. The third term in (38) is reminiscent of the correlation functions discussed in the previous section, except that this term is a product of the structure amplitudes $|F^{II}|$ and $|F^{III}|$, instead of the intensities. We know that, as a first approximation, the positions of the peaks in the syntheses $|F^{III}|$ and $|F^{II}|$ correspond to the Pattersons $|F^{III}|^2$ and $|F^{II}|^2$, respectively, and hence the product $|F^{III}|\,|F^{II}|$ would be closely similar to the correlation function $|F^{II}|^2\,|F^{III}|^2$.

We can show that the synthesis $|F^{II}|\,|F^{III}|$ has strong peaks at positions $\mathbf{r}_{Pi}^{II} - \mathbf{r}_{Pj}^{III}$ by making use of the results of Ch. 2. From Table 2 of that chapter, the structure $|F^{II}|$ has an origin peak of strength S_N^{II} and nonorigin peaks at $(\mathbf{r}_{Ni}^{II} - \mathbf{r}_{Nj}^{II})$ of strengths $f_{Ni}^{II}f_{Nj}^{II}/2S_N^{II}$. Similar results hold for the structure $|F^{III}|$. The synthesis $|F^{II}|\,|F^{III}|$ is the convolution of the structures $|F^{II}|$ and $|F^{III}|$, and it is readily verified that the peak corresponding to the correlation vector $(\mathbf{r}_{Pi}^{II} - \mathbf{r}_{Pj}^{III})$ arises mainly from the interactions between the vectors $(\mathbf{r}_{Pi}^{II} - \mathbf{r}_{Qk})$ in structure $|F^{II}|$ and the vectors $(\mathbf{r}_{Qk} - \mathbf{r}_{Pj}^{III})$ in structure $|F^{III}|$. It follows that the strength of the correlation peaks at this location is equal to $f_{Pi}^{II} f_{Pj}^{III} S_Q^2 / 4 S_N^I S_N^{II}$. It is interesting that the strength of the correlation peaks is directly proportional to S_Q^2 and hence will be larger, the larger is the number of the common Q atoms in the two structures.

The main interest in (38) is that, since the product $|F^{II}||F^{III}|$ occurs in a negative term, the interaction peaks between the atoms P_i^{II} and P_j^{III} will appear with *negative strength* in this synthesis. This has been found to work well, as tested by Rossmann.[41] Another function[19-21,54] containing only the moduli that can be used as a correlation function is

$$(|F^{II}| - |F^I|)(|F^{III}| - |F^I|). \qquad (39)$$

This will obviously contain the interaction peaks between the heavy atoms with positive strength, for, on expansion, (39) is seen to contain the product $|F^{II}||F^{III}|$ with a positive sign.

It should be mentioned that the above functions involving the moduli, i.e., the first power of the amplitudes, are in general good approximations only in limiting situations when $F_P \ll F_N$. Fortunately this is true especially in protein crystallography, where the heavy atoms in the derivatives of a parent protein make only small mean changes in the intensities of reflections ($\simeq 10\%$). For more general situations, formulas involving the intensities are preferable.

We have considered in this chapter mainly the problems involved when isomorphous data are available and how best they can be used. In Ch. 11 we shall see that when anomalous scattering of x-rays is present, a parallel method of analysis is possible, which however has features complementary to those of the isomorphous case. In fact, a combination of the data from isomorphous crystals and anomalous dispersion offers a powerful tool in the unique solution of the phase problem, and this will be considered in Ch. 13.

Examples of the determination of the replaceable-atom positions

We give here a few illustrations of the different syntheses described earlier for the location of the heavy atoms, as well as of their relative positions.

The principle of maximum superposition on the DP was tested by Sarma and Srinivasan[44] on a semi-hypothetical pair of isomorphous compounds. The pair chosen was tyrosine hydrochloride and hydrobromide. However, the chlorine and bromine atoms were removed from their respective structures and an oxygen atom was assumed at the position of the bromine atom in the hydrobromide, while the tyrosine molecule, as found in the hydrochloride, was retained in the other crystal, with no atom in the position of the chlorine.

In this way, a pair was obtained, one of which contained an additional atom of almost the same scattering power as the rest of the structure, and also, possible differences in the coordinates of the tyrosine molecule in the two structures were simulated. The difference Patterson of the b axis projection of the above pair is shown in Fig. 6. (This is a centrosymmetric projection,

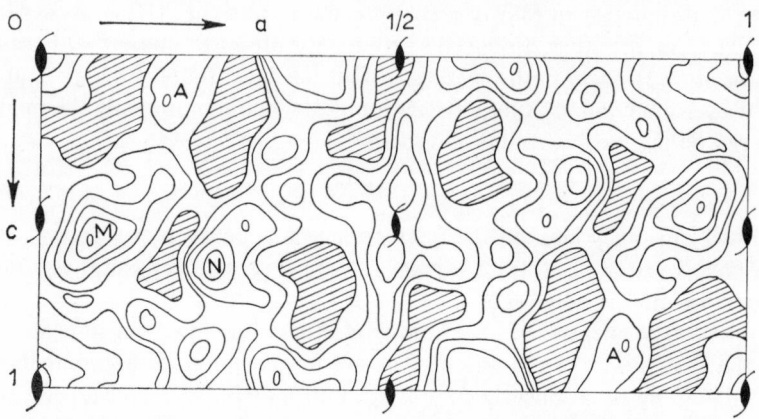

Fig. 6. The ac projection of the sharpened difference Patterson of the structures tyrosine and tyrosine + oxygen. Shaded regions correspond to negative values.

the space group being $P2_1$.) The position of the single interaction peak between the additional oxygen atoms is marked by A in this figure. This is known in the present case, since we are dealing with a known model. It will be noticed that the strength at the terminus of the P–P vector is of the same order of magnitude, if not smaller than some of the other peaks in the diagram, such as those marked by M, N. Thus, with a mean contribution to the intensity of about 10 % by the P atoms, with possible errors in coordinates of the molecules in the two crystals, the location of the heavy-atom peak in the DP would appear to be quite difficult. In fact, in connection with studies on proteins, Blow[1] has pointed out that background peaks in the DP of horse hemoglobin were up to $2\frac{1}{2}$ times the strength of peaks due to the heavy atoms. Another example of the failure of the DP to reveal the additional P atom vectors is the case of caffeine and theophylline.[55]

Principle of maximum superposition In the example of the tests with the tyrosine hydrohalides, however, the replaceable-atom interactions could be detected by making use of the principle of maximum superposition[44] and calculating the cumulative minimum function $C(t)$. Vector shifts were made to the point A, as well as to two other arbitrary peaks marked M and N, in

Table 2   Values of the cumulative minimum function corresponding to the various vector shifts in Fig. 6[†]

x z	4/60	6/60	8/60	10/60	12/60	14/60
4/60				4.89		
6/60			4.26 (2471)	5.02 (2608) A	4.65 (2573)	
7/60			4.38	5.06 (2624)	4.47	
8/60			4.51	5.03	4.32	
10/60				4.93		
34/60	M 4.05					
40/60						N 3.67 (2401)

[†] The coordinates are given with respect to the top left corner of Fig. 6 as origin. The values are scaled down by a factor of 10^4. Those given in brackets correspond to the cumulative product function (after Sarma and Srinivasan [44]).

Fig. 6 and the minimum function was obtained in each case. The values of the minimum function at grid points all over the entire cell were summed up and this gave the magnitudes of the superposition, corresponding to each vector shift. These are given in Table 2 for a number of points around the peak A and also corresponding to the two peaks M and N. It may be noticed that even though the actual peak strengths at M and N in the DP are larger than at A, the cumulative minimum function exhibits a much higher peak at A compared to the values at M and N. The above was tried only in projection. In three dimensions, therefore, we could expect it to work even better.

MD-squared synthesis   The MD-squared synthesis was computed for the pair considered above and this is shown in Fig. 7. The position where a peak is expected corresponding to the $P–P$ vector is marked by a cross in Fig. 7. Although there is a peak of reasonable strength at the expected position, the map does not seem to be fully satisfactory, as there are other broad peaks and regions of positive strength of comparable magnitude. The synthesis, however, seems to have worked well in the practical case of proteins in centrosymmetric projections.[14] A clear demonstration of its success in a

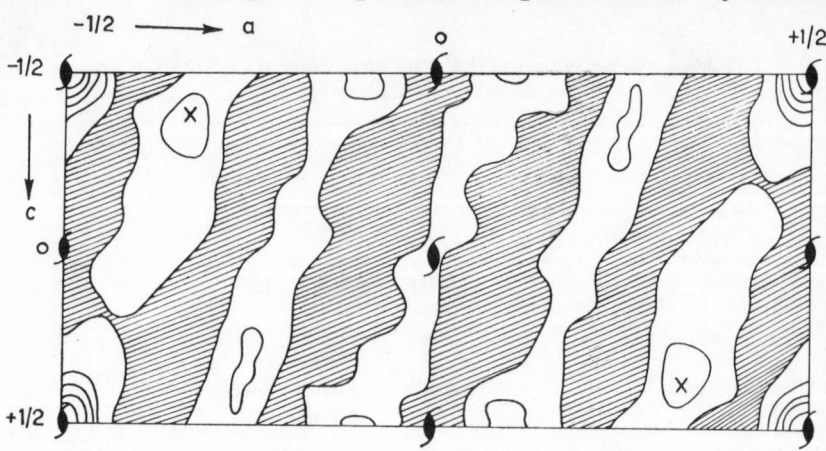

Fig. 7. The *MD*-squared synthesis for the same pair as in Fig. 6. The correct position of the O–O interaction is marked by a cross.

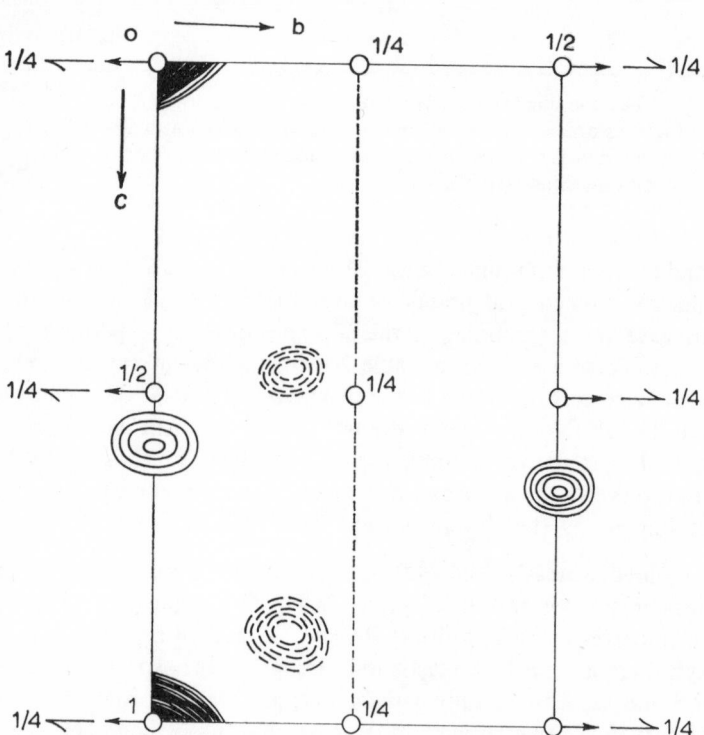

Fig. 8. Composite view of the *MD*-squared synthesis for the two derivatives HgAc$_2$ and PCMB of haemoglobin. $(|F_{\text{HgAc}_2}| - |F_{\text{PCMB}}|)^2$ synthesis looking down through one quarter of the cell. Dashed lines are negative contours.

noncentrosymmetric case was first given by Rossmann[41] with hemoglobin He used the function $(|F^{III}| - |F^{II}|)^2$ where $|F^{III}|$ and $|F^{II}|$ are the structure. amplitudes of two derivatives of the parent protein. The resultant map (Fig. 8) not only gave the heavy-atom peaks of the two individual structures but also their relative positions by a negative peak, as derived theoretically earlier in this section. Other examples of the successful use of the MD-squared synthesis are in the solution of the structures of ribonuclease[19] and lysozyme.[30]

The weighted form of MD-squared synthesis was tested by Kalyanaraman and Srinivasan[17] and is illustrated in Fig. 9a. The mean contribution to the intensity by the heavy atom was taken to be about 11% and it may be seen that there is improvement in the weighted form as compared with the ordinary MD-squared synthesis which is given in Fig. 9b for comparison. In particular

(a)

Fig. 9. (a) Modified MD-squared synthesis;

(b)

Fig. 9. (*b*) Simple *MD*-squared synthesis for a hypothe-
itcal pair of (addition-type) isomorphous crystals. Mean
contribution of the heavy-atom to the intensity is 11%.
Negative regions are shaded. True single and double peaks
are marked as X and ●, respectively. Note the suppression
of spurious details in (*a*).

the spurious peak *A* seen in the *MD*-squared is practically absent in the
weighted one.

The *MD*-squared synthesis has been tried in another entirely different
situation, making use of the simulation of isomorphism by the use of different
wavelengths (discussed in Ch. 9). It was calculated [39,40] for the case of
$KMnO_4$ using the intensities measured for two different wavelengths. Since
the isomorphism is exact, even the small change in the real part of the
scattering factor was sufficient to reveal the replaceable atom interactions
(Fig. 10).

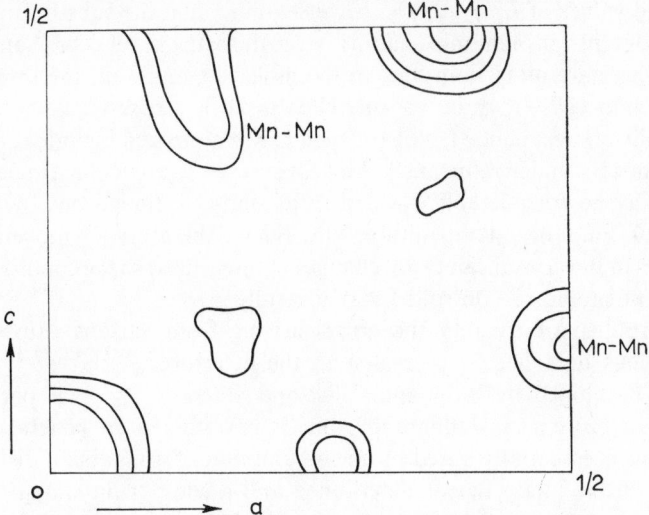

Fig. 10. The ac projection of MD-squared synthesis for $KMnO_4$ with measurements at two wavelengths. Synthesis was done with coefficients $(|F_{CuK\alpha}| - |F_{FeK\alpha}|)^2$. (After Ramaseshan et al.[40])

Practical aspects of treatment of isomorphous data

The various formulas and methods discussed earlier for isomorphous crystals are all based on the assumption that true isomorphism exists between the pair of crystals for which the calculations are made. So also the formulas involve $|F^I|$, $|F^{II}|$, $|F^{III}|$, which are the true amplitudes of the relevant crystals, whereas in practice we can use only the observed quantities which are subject to errors (not only observational, but also arising from errors in scaling of the individual data). The other quantities that enter the calculations, such as F_P^{II}, F_P^{III}, δF_P, are calculated values and, if the positions of the replaceable atoms have been correctly determined, the possible errors in them may be taken to be of less importance compared to the errors in $|F^I|$, $|F^{II}|$, $|F^{III}|$.

However, in general, all these quantities are subject to errors. True isomorphism rarely exists in practice between a pair of crystals. Substitution of a group of atoms by another, as in the replacement-type isomorphism, or addition of heavy atoms or groups of atoms to a parent molecule, is often likely to disturb the environment and, hence, the coordinates of the rest of the atoms will not be identical in the two crystals.

The above problems assume considerably greater importance when we come to complex structures such as proteins. For example, apart from the usual types of errors mentioned above, we cannot also be certain in this case of the calculated values of F_P^{II}, F_P^{III}, δF_P, because quite often the substitution of the heavy atom into protein molecules is never the same in all cells. Some of the sites (if there are more than one) in the molecule, and even the same site in adjacent unit cells, may be vacant. Thus in such a case we have to assign statistically an occupancy factor for each heavy atom and include it as one of the parameters under refinement. Moreover, even a particular single protein crystal may be considerably affected in its unit-cell dimensions by the conditions of humidity, temperature, etc. Such "breathing" movements of molecules in the crystal can cause changes in intensities; so the condition of the experiment should be controlled very carefully.

The problem of treating the errors arising from various causes in the isomorphous data has been treated in the literature.[2,9–11,24,27,28,35,47–53] The one developed by Blow and Crick and others[11,24,27,43] is particularly relevant to protein crystallography, but is nevertheless of general applicability. This is essentially based on the assumption of a Gaussian distribution of errors in the phase that is determined and a consequent statistical treatment of the data. On the other hand, methods have been developed in the authors' laboratory for an assessment from a statistical study of their intensity data of the degree of isomorphism between a given pair of crystals. We shall outline these different studies in the following sections.

The effect of errors in the data on phase-angle calculation

Crick and Magdoff[9] have analyzed the effect of various errors, in particular, lack of isomorphism, on the intensities. They showed that the lack of isomorphism affects, in an increasing manner, higher-angle reflections. They estimated the average change in intensity due to the addition of heavy atoms to a protein crystal, and also derived formulas for the change in intensity caused by small amounts of coordinate shifts of molecules, as well as small rotations and translations due to alterations in lattice parameters in the two crystals. They concluded that small molecular shifts would interfere with the isomorphous replacement method at higher values of $\mathbf{r}^*$, but that the formulas could be used at lower resolutions, provided some caution is exercised. Such a result was to be expected in view of an earlier analysis in another context by Luzzati,[23] the extension of whose results to isomorphous crystals has been made more recently.[35,48–53]

The problem of the effect of various errors in isomorphous data on the phase and their treatment was considered by Blow and Crick[2] and has been largely extended by later workers in handling the data of a series of isomorphous crystals,[11] as well as of the inclusion of anomalous dispersion data.[24,27] Their approach can be understood in the following way.

Consider the simple case of a heavy-atom structure in which the heavy-atom positions are known exactly, so that the structure factor F_P of the heavy atoms is known. Let us also assume that the true value of the structure amplitude $|F_N|$ of the entire structure is also known. The best that could be said about the unknown phase angle ϕ_N under these conditions is to assign a probability distribution for it. The uncertainty in the phase ϕ_N is obviously due to the lack of knowledge of the vector F_Q, and since we know the statistical distribution of the magnitude $|F_Q|$ from Wilson's statistics, we can work out what should be the phase distribution of ϕ_N. Purely from a priori considerations, it is obvious that the distribution should be symmetrical with respect to the phase of the heavy-atom group. In fact, the distribution is given conveniently for the angle $\theta = (\phi_N - \phi_P)$, which is the deviation of the phase ϕ_N from ϕ_P.

Consider now a pair of isomorphous crystals of the addition type, given by F_Q and $F_N = F_Q + F_P$. If the true values are available, we can obtain the phase angle (for convenience, say ϕ_Q of F_Q, the invariant part of the two isomorphous structures), except for a two-fold ambiguity, (4) of Ch. 9. In terms of a probability distribution, the two values obtained may be considered as two delta functions corresponding to the two values which are calculated. On the other hand, in real practice the calculated values are obtained from a set of observed structure amplitudes and calculated F_P values, which are subject to various errors as described above. Thus the calculated phase is no longer an exact one and we have to assign a probability distribution to the phase, similar to what was discussed in Ch. 8 for the heavy-atom case. Obviously the distribution in the present case is essentially determined by the various types of errors (lack of isomorphism, errors of observation, errors in scaling, etc.). Since these errors may be considered to be random in general, we could make a simple assumption that the phase error has a Gaussian distribution. In the method of Blow and Crick, it is first assumed that the observed value of $|F_Q|$ is the true value and that whatever errors are present may be put into the errors in the observed value of $|F_N|$. The calculated value of $|F_N|$, which is obtained by adding F_Q and F_P, will differ, in general, from the "observed" value, the difference

$$|F_N^{obs}| - |(F_P + F_Q)| = \varepsilon(\phi_Q) \tag{40}$$

being called the lack of closure. If we call $F_P + F_Q = D$, D is obtained from the formula

$$D^2 = |F_Q|^2 + |F_P|^2 + 2|F_Q||F_P|\cos(\phi_P - \phi_Q). \tag{41}$$

The term "lack of closure" is obvious, since it arises from the noncoincidence of the calculated third vector $F_N(= F_P + F_Q = D)$ with the vector F_N^{obs} (Fig. 11). It may be noticed that the quantity ε is shown as a function of ϕ_Q, since with every value of ϕ_Q associated with the phase of F_Q, we could calculate

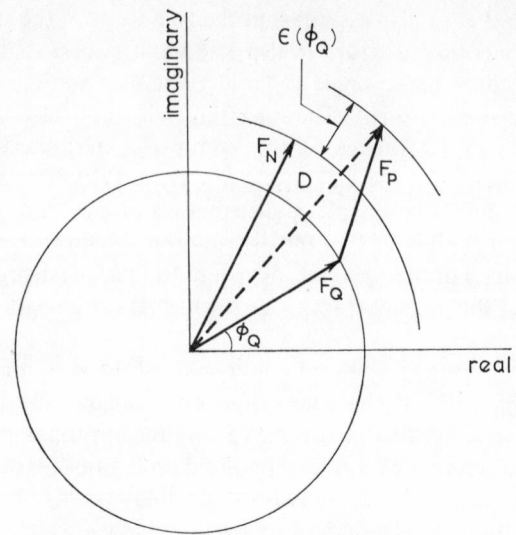

Fig. 11. Lack-of-closure error in isomorphous data. For a given F_Q, different values of ϕ_Q are associated with different lack-of-closure error $\epsilon(\phi_Q)$ obtained by the triangle solution.

such a value from (41). The probability distribution $P(\phi_Q)\,d\phi_Q$ is now assumed to be given by

$$P(\phi_Q)\,d\phi_Q = C \exp\left[\frac{-\epsilon^2(\phi_Q)}{2E^2}\right], \tag{42}$$

where C is a normalizing constant and E^2 is the variance of the distribution $\epsilon(\phi_Q)$. We shall consider a little later how the expression (42) can be set up in a practical case. Assuming that the distribution $P(\phi_Q)$ is known, the best Fourier synthesis can be shown to be one which uses the amplitude $|F_Q|$ and the expectation value of the phase $\langle\exp(i\phi_Q)\rangle$ as coefficients. Thus consider the error in electron density arising from errors in a given reflection and its conjugate. If the structure factor used in the synthesis is F_Q^0 and its true value is F_Q^t, then the mean square error in the electron density for this pair (i.e., the mean square value of the difference Fourier, arising from this pair) is

$$\langle(\Delta\rho)^2\rangle = \frac{2}{V^2}(F_Q^0 - F_Q^t)^2. \tag{43}$$

Considering the vector F_Q^t, which can be written as

$$F_Q^t = |F_Q|\,\mathbf{u}(\phi_Q), \tag{44}$$

where $\mathbf{u}$ is the unit vector in the Argand diagram in the direction ϕ_Q, one can now assign a distribution for the phase angle ϕ_Q. Equation (43) then becomes

$$\langle(\Delta\rho)^2\rangle = \frac{2}{V^2} \int_{\mathbf{u}(\phi_Q)} [F_Q^0 - |F_Q|\,\mathbf{u}(\phi_Q)]^2 P(\phi_Q)\,d\mathbf{u}(\phi_Q). \qquad (45)$$

If this is to be a minimum, we readily get from (45)

$$F_Q^0(\text{best}) = \frac{\displaystyle\int_{\mathbf{u}(\phi_Q)} |F_Q|\,P(\phi_Q)\mathbf{u}(\phi_Q)\,d\mathbf{u}(\phi_Q)}{\displaystyle\int_{\mathbf{u}(\phi_Q)} P(\phi_Q)\,d\mathbf{u}(\phi_Q)}, \qquad (46a)$$

$$= m\,|F_Q|\,\exp\,[i\phi_Q(\text{best})], \qquad (46b)$$

say, where $F_Q^0(\text{best})$ stands for the "best Fourier" in the sense that $(\Delta\rho)^2$ is minimized. This procedure may be looked upon as an extension of the simple method of using both the phases obtained in the isomorphous case, when all the quantities are known exactly. The use of both phases (when there are no errors at all) is equivalent to taking the "mean" of the two values represented by two delta functions. When there are errors, we have, instead of two delta functions, a continuous distribution, and we replace the mean of the two by the expectation value of the phase factor $\exp(i\phi_Q)$ over the range of variation of ϕ_Q.

This can be pictured in another convenient way. Thus in Fig. 12 a unit circle is drawn with the origin of the Argand diagram as the center. It defines the locus of the terminus of the unit vector $\mathbf{u}(\phi_Q)$, which is equal to $\exp(i\phi_Q)$. For any given ϕ_Q, the distribution $P(\phi_Q)\,d\phi_Q$ can be conveniently represented by a length $P(\phi_Q)$ along the direction defined by $\mathbf{u}(\phi_Q)$ with the circumference of the circle as base. For a given distribution $P(\phi_Q)\,d\phi_Q$, as given by (42), there are, in general, two maxima in the graph so obtained, centered at the two true solutions of the phase angle for the given pair of isomorphous crystals. The phase to be used is the expectation value of $\exp(i\phi_Q)$, which is given by:

$$\langle\exp(i\phi_Q)\rangle = \frac{\int \exp(i\phi_Q)P(\phi_Q)\,d\phi_Q}{\int \exp(i\phi_Q)\,d\phi_Q}. \qquad (47)$$

If we consider the unit phase circle as a loop of wire of variable density, the density at each point being proportional to the probability of the phase ϕ_Q given by (42), then (47) is equivalent to finding the centroid of the distribution, or the center of gravity of the wire. This obviously lies within the circle and is given by the point C at the terminus of the vector $\mathbf{m}$, with the polar coordinates (m, ϕ_B). (The magnitude of $\mathbf{m}$ will be large, the sharper is the distribution $P(\phi_Q)$. In the limit, when $P(\phi_Q)$ has a flat uniform distribution,

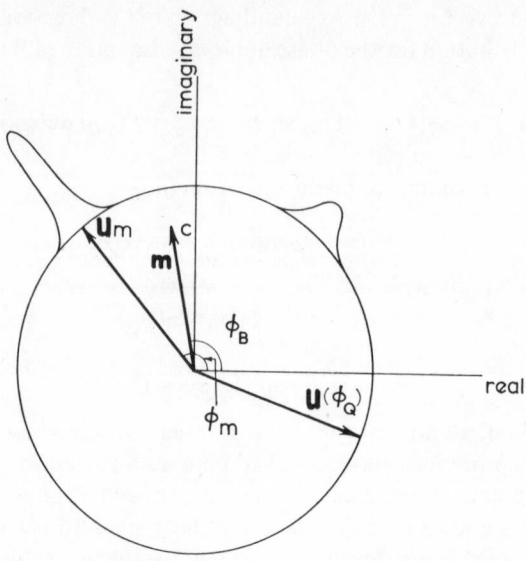

Fig. 12. Unit phase circle with phase probability $P(\phi_Q)$ plotted with circle as base. $\mathbf{u}_m$ corresponds to the most probable phase and $\mathbf{m}$ corresponds to the centroid. The distribution has been purposely made asymmetric so that it is general. A single pair leads to a symmetric distribution on either side of the centroid. (After Dickerson et al.[11])

i.e., a constant for all ϕ_Q, $\mathbf{m} \to 0$; this corresponds to a large uncertainty in the phase ϕ_Q.) The Fourier synthesis performed on this basis, namely the one corresponding to the centroid $\mathbf{m}$, is termed the "best Fourier." It is equivalent to weighting each $|F_Q|$ by the factor m and using the phase ϕ_Q (best). It may be noticed that the above Fourier is in general different from the one using the most probable phase. The latter corresponds to using the phase of the maximum in the probability curve ($\mathbf{u}_m$ in Fig. 12). The quantity m is called the "figure of merit." It may be readily shown that it is the expectation value of the cosine of the error in the phase angle ϕ_Q. It may also be termed *the phase correlation*, a term used by Srinivasan and Chandrasekaran[47] in another context.

The above treatment could be extended in general to a series of isomorphous compounds.[11] A given pair yields, in general, a "bimodal" distribution. Addition of one further ismorphous compound, for example, may be taken to be an independent source of information about the phase distribution $P(\phi_Q)$ and hence the total probability can be taken to be the product of the two individual ones. In general, for a given series of compounds, we could write

$$P(\phi_Q) = \prod_{j=1}^{n} P_j(\phi_Q) = \exp - \sum_{j=1}^{n} \frac{\varepsilon_j^2(\phi_Q)}{2E_j^2}, \tag{48}$$

The bimodal distribution will, in general, lose its symmetry and one of the peaks will tend to grow and the others to decrease. It may be noticed that, according to the above treatment, it is strictly incorrect to "pick out" the common phase from a series, but we must use the centroid of the resultant phase distribution.

In practice, for a given pair, the phase distribution $P(\phi_Q)$ may be obtained as follows. If the crystal has a centrosymmetric projection, then we can assume that

$$E^2 = \langle(|F_N| - |F_Q| - |F_P|)^2\rangle \tag{49}$$

for that zone and this estimated value can then be used in the general phase distribution (42) for all the other reflections as well.

The above treatment has worked satisfactorily in protein-structure determinations. There are, however, two features that are not satisfactory and need improvement. The first is that the value of $|F_Q|$ observed is treated to be the "true" value without any error and all errors are taken to reside in the data of the heavy-atom derivative F_N. Cullis et al.[10] have formulated a slightly different method to overcome this defect. We shall not, however, go into the details of this here. The second is the assumption that E_j^2 assumed in the distribution (42) is the same as the one obtained for a centrosymmetric zone. In the absence of any other alternative information, this would appear to be the only course to be taken. However, a better approach would be to calculate the statistical parameters which define the degree of isomorphism from the experimental data and use them in the phase distribution.[47] One of these parameters, namely σ_A (see the next subsection) can be used to characterize the phase distribution. These parameters can be obtained by calculating the normalized reliability indices and other calculable parameters using the entire data of observed and calculated structure factors, instead of those of a centric zone alone. However, these ideas are applicable only on the basis that errors of observation affect the theory in a way similar to (or at least approximately as) lack of isomorphism which is treated by these authors.

The method of Blow and Crick has been extended to include the case of anomalous dispersion data,[24,27] and also in the general problem of refinement of protein phases by iterative methods.

Statistical tests for isomorphism

In a series of papers from the authors' laboratory,[28,35,46-53] the problem of the statistical distribution of x-ray intensities of a pair of crystals was considered, and this is particularly applicable to the case of isomorphous crystals of the addition type. The distributions associated with the structure amplitudes of a pair of crystals (such as the difference, product, quotient, etc.), were theoretically examined and these have yielded statistical criteria which can be used to assess the degree of isomorphism of the pair. Statistical parameters in the nature of reliability indices have been developed and they involve the use of the observed intensities and amplitudes of the two crystals. A detailed analysis of the problem shows that the errors in the coordinates of the common atoms in the two crystals affect progressively higher-angle reflections to a larger extent. Making use of the parameters and tests, it is possible to obtain an estimate of the mean error in the coordinates of the common atoms.

The above treatment includes strictly only possible errors in the coordinates of the common part, while otherwise the true structure amplitudes are assumed to be available. It would however appear to be possible to extend the use of these statistical parameters even for the case when the observed structure amplitudes suffer from observational errors. The nature of the mathematical formulas suggest that, since the observational errors are expected to be random, namely having a Gaussian type of distribution, the interpretation of the final results would still be valid at least as an approximation. We shall not discuss these here and the original papers may be referred to for more details.

References

[1] D. M. Blow. *The structure of haemoglobin VII. Determination of phase angles in the noncentrosymmetric* (1 0 0) *zone.* Proc. Roy. Soc. **A247** (1958) 302–336.

[2] D. M. Blow and F. H. C. Crick. *The treatment of errors in the isomorphous replacement method.* Acta Crystallogr. **12** (1959) 794–802.

[3] D. M. Blow and M. G. Rossmann. *Single isomorphous replacement method.* Acta Crystallogr. **14** (1961) 1195–1202.

[4] G. Bodo, H. M. Dintzis, J. C. Kendrew, and H. W. Wyckoff. *The crystal structure of myoglobin V. A low-resolution three-dimensional Fourier synthesis of sperm-whale myoglobin crystals.* Proc. Roy. Soc. **A253** (1959) 70–102.

[5] C. Bokhoven, J. C. Schoone, and J. M. Bijvoet. *The Fourier synthesis of the crystal structure of strychnine sulphate pentahydrate.* Acta Crystallogr. **4** (1951) 275–280.

[6] W. L. Bragg. *The determination of the coordinates of heavy atoms in protein crystals.* Acta Crystallogr. **11** (1958) 70–75.

[7] M. J. Buerger. *A new Fourier series technique for crystal structure determination.* Proc. Nat. Acad. Sci. U.S. **28** (1942) 281–285.

[8] M. J. Buerger. *Vector space and its application in crystal structure determination.* (Wiley, New York, 1959) pp. 66 et. seq.

[9] F. H. C. Crick and B. S. Magdoff. *The theory of the method of isomorphous replacement for protein crystals I.* Acta Crystallogr. **9** (1956) 901–908.

10 A. F. Cullis, H. Muirhead, M. F. Perutz, M. G. Rossmann, and A. C. T. North. *The structure of haemoglobin VIII. A three-dimensional Fourier synthesis at 5.5 A resolution: Determination of the phase angles.* Proc. Roy. Soc. A265 (1962) 15–38.

11 R. E. Dickerson, J. C. Kendrew, and B. E. Strandberg. *The crystal structure of myoglobin: Phase determination to a resolution of 2A by the method of isomorphous replacement.* Acta Crystallogr. 14 (1961) 1188–1195.

12 Gabrielle Donnay and M. J. Buerger. *The determination of the crystal structure of tourmaline.* Acta Crystallogr. 3 (1950) 379–388.

13 Alfred J. Frueh. *An extension of " the difference Patterson" to facilitate solution of order-disorder problems.* Acta Crystallogr. 6 (1953) 454–456.

14 D. W. Green, V. M. Ingram, and M. F. Perutz. *The structure of haemoglobin. VI. Sign determination by the isomorphous replacement method.* Proc. Roy. Soc. A225 (1954) 287–307.

15 D. Harker. *The determination of phases of structure factors of noncentrosymmetric crystals by the method of double isomorphous replacement.* Acta Crystallogr. 9 (1956) 1–9.

16 W. Hoppe. *Die Bestummung genauer schweratom Parameter in isomorphen azentrischen Kristallen.* Acta Crystallogr. 12 (1959) 665–674.

17 A. R. Kalyanaraman and R. Srinivasan. *The best Patterson function for the determination of heavy-atom positions using isomorphous crystals.* Z. Kristallogr. 126 (1968) 262–267.

18 G. Kartha. *Isomorphous replacement method in noncentrosymmetric structures.* Acta Crystallogr. 14 (1961) 680–686.

19 G. Kartha, J. Bello, D. Harker, and F. E. DeJarnette. *The structure of ribonuclease II. The positions of the heavy atoms in five " dyed" crystals.* In "Aspects of protein structure" (Ed.). G. N. Ramachandran, Academic Press, London (1963) 13–22.

20 G. Kartha and R. Parthasarathy. *Combination of multiple isomorphous replacement and anomalous dispersion data for protein structure determination I. Determination of heavy atom positions in protein derivatives.* Acta Crystallogr. 18 (1965) 745–749.

21 G. Kartha and R. Parthasarathy. *Combination of multiple isomorphous replacement and anomalous dispersion data for protein structure determination II. Correlation of heavy-atom positions in different crystals.* Acta Crystallogr. 18 (1965) 749–753.

22 G. Kartha and G. N. Ramachandran. *Applications of the difference Patterson technique in structure analysis.* Acta Crystallogr. 8 (1955) 195–199.

23 V. Luzzati. *Traitment statistique des erreurs dans la determination des structures crystallines.* Acta Crystallogr. 5 (1952) 802–810.

24 B. W. Mathews. *The extension of isomorphous replacement method to include anomalous scattering measurements.* Acta Crystallogr. 20 (1966) 82–86.

25 S. K. Mazumdar. *A test of the β-isomorphous syntheses.* Indian J. Pure Appl. Phys. 3 (1965) 411–413.

26 S. K. Mazumdar and R. Srinivasan. *The crystal structure of L-arginine hydrobromide.* Z. Kristallogr. 123 (1966) 186–205.

[27] A. C. T. North. *The combination of isomorphous replacement and anomalous scattering data in phase determination of noncentrosymmetric reflexions.* Acta Crystallogr. **18** (1965) 212–216.

[28] S. Parthasarathy and R. Srinivasan. *Probability distributions connected with intensities of two related crystals.* Indian J. Pure Appl. Phys. **5** (1967) 502–510.

[29] M. F. Perutz. *Isomorphous replacement and phase determination in noncentrosymmetric space groups.* Acta Crystallogr. **9** (1956) 867–873.

[30] D. C. Phillips. *Advances in protein crystallography* in "Advances in Structural Research" (Ed.) Brill and Mason. Springer Verlag. (1966).

[31] G. N. Ramachandran. *Fourier methods in structure analysis.* Proceedings of the Physics Conference and Symposium on Solid State Physics, Bangalore (1960) 131–150.

[32] G. N. Ramachandran. *Fourier syntheses for partially known structures.* In "Advanced methods of crystallography" (Ed.) G. N. Ramachandran, Academic Press, London (1963) 25–65.

[33] G. N. Ramachandran and R. R. Ayyar. *Fourier syntheses for feeding in isomorphous replacement and anomalous dispersion data.* In "Crystallography and crystal perfection". Ed. G. N. Ramachandran. Academic Press, London (1963) 25–41.

[34] G. N. Ramachandran and S. Raman. *Syntheses for the deconvolution of the Patterson function. Part I. General principles.* Acta Crystallogr. **12** (1959) 957–964.

[35] G. N. Ramachandran, R. Srinivasan, and V. Raghupathy Sarma. *Probability distribution connected with structure amplitudes of two related crystals. Part I. Probability distribution of the difference.* Acta Crystallogr. **16** (1963) 662–666.

[36] S. Raman. *Syntheses for the deconvolution of the Patterson function. Part II. Detailed theory for noncentrosymmetric crystals.* Acta Crystallogr. **12** (1959) 964–975.

[37] S. Raman. *Syntheses for the deconvolution of the Patterson function. Part III. Theory for centrosymmetric crystals.* Acta Crystallogr. **14** (1961) 148–150.

[38] S. Raman and W. N. Lipscomb. *Two classes of functions for the location of heavy-atoms and for solution of crystal structures.* Z. Kristallogr. **116** (1961) 314–327.

[39] S. Ramaseshan and K. Venkatesan. *The use of anomalous scattering without phase change in crystal structure analysis.* Curr. Sci. (India) **26** (1957) 352–353.

[40] S. Ramaseshan, K. Venkatesan, and N. V. Mani. *The use of anomalous scattering for the determination of crystal structures* $KMnO_4$. Proc. Ind. Acad. Sci. **A46** (1957) 95–111

[41] M. G. Rossmann. *The accurate determination of the position and shape of heavy-atom replacement groups in proteins.* Acta Crystallogr. **13** (1960) 221–226.

[42] M. G. Rossmann and D. M. Blow. *Single isomorphous replacement method. A correction.* Acta Crystallogr. **15** (1962) 1060.

[43] M. G. Rossmann and D. M. Blow. *The refinement of structures partially determined by the isomorphous replacement method.* Acta Crystallogr. **14** (1961) 641–647.

[44] V. Raghupathy Sarma and R. Srinivasan. *Principle of maximum superposition. A method for determining the positions of replaceable atoms in isomorphous crystals.* Acta Crystallogr. **15** (1962) 457–460.

[45] V. Sasisekharan. *An application of the difference Patterson method.* Proc. Ind. Acad. Sci. **43A** (1956) 224–230.

[46] R. Srinivasan. *Weighting function in crystal structure analysis.* Z. Kristallogr. **126** (1968) 175–181.

[47] R. Srinivasan and R. Chandrasekaran. *Correlation functions connected with structure factors and their application to the case of the observed and calculated structure factors.* Ind. J. Pure Appl. Phys. **4** (1966) 178–186.

[48] R. Srinivasan, V. Raghupathy Sarma, and G. N. Ramachandran. *Probability distribution connected with structure amplitudes of two related crystals. Part II. Probability distribution of the product.* Acta Crystallogr. **16** (1963) 1151–1156.

[49] R. Srinivasan, V. Raghupathy Sarma, and G. N. Ramachandran. *Statistical tests for isomorphism.* In " Crystallography and crystal perfection " (Ed.) G. N. Ramachandran, Academic Press, London (1963) 85–98.

[50] R. Srinivasan and G. N. Ramachandran. *Probability distribution connected with structure amplitudes of two related crystals. Part IV. Distribution of the normalised difference.* Acta Crystallogr. **19** (1965) 1003–1007.

[51] R. Srinivasan and G. N. Ramachandran. *Probability distribution connected with structure amplitudes of two related crystals. Part V. The effect of errors in the atomic coordinates on the distribution of the observed and calculated structure factors.* Acta Crystallogr. **19** (1965) 1008–1014.

[52] R. Srinivasan and G. N. Ramachandran. *Probability distribution connected with structure amplitudes of two related crystals. Part VI. On the significance of the parameter σ_A.* Acta Crystallogr. **20** (1966) 570–571.

[53] R. Srinivasan, E. Subramanian, and G. N. Ramachandran. *Probability distribution connected with structure amplitudes of two related crystals. Part III. Probability distribution of the quotient.* Acta Crystallogr. **17** (1964) 1010–1014.

[54] L. K. Steinrauf. *Two Fourier functions for use in protein crystallography.* Acta Crystallogr. **16** (1963) 317–318.

[55] D. June Sutor. *Isomorphous replacement applied to molecules containing like atoms.* Acta Crystallogr. **9** (1956) 969–970.

11

Anomalous-dispersion effect: general aspects

The anomalous-dispersion effect for x-rays

So far, we have been considering how the partial information about a structure, in the form of the nature and positions of some of its atoms, can be utilized for developing the rest of the structure. We also considered the situation of isomorphism, which could be taken to be an aspect of partial information, in the sense that such partial information is available from two sources in the form of two isomorphous crystals, instead of only one. There is another situation in which similar extra information becomes available, and this, in fact, offers an equally powerful method of analysis. This arises if some of the atoms in the structure scatter x-rays anomalously. In this chapter and the next, we shall be concerned mainly with the application of this effect for the solution of crystal structures.

For a detailed account of the anomalous-dispersion effect itself, reference may be made to standard books such as that by James.[15] The essential facts about this phenomenon can be summarized as follows. When the wavelength of the incident x-rays is close to an absorption edge of an atom, the atomic scattering factor is no longer real, but becomes a complex quantity, given by

$$f = f_0 + \delta f' + i\,\Delta f'', \tag{1}$$

where $\delta f'$ and $\Delta f''$ are the dispersion and absorption corrections to the normal scattering factor of the atom.‡ The atomic scattering factor f measures, in

‡ We use small δ for the dispersion correction, instead of Δ commonly used, since its application leads essentially to an effect analogous to isomorphous replacement, for which the change in scattering factor is denoted by δf_P in this monograph.

166

units of the scattering from a Thomson free electron, the coherent scattering amplitude for an atom composed of free electrons. However, this is actually not the case, and none of the electrons in any atom is really free. The fact that the electrons are bound to the nucleus necessarily brings in resonance phenomena, exhibited in the form of absorption edges for x-rays in the atom. The theoretical derivation of the scattering factor for any atom involves, in general, terms dependent on the frequency of the incident x-rays. The normal amplitude f_0 is only an approximation, valid for incident wavelengths far removed from absorption edges. This is a good enough approximation for most cases of light atoms, whose absorption edges are in the soft x-ray region; however, the deviation from f_0 becomes quite significant for atoms with larger atomic numbers. Obviously the absorption edges which play an important role correspond to the innermost electronic shells, (K, L, principally), which are more tightly bound to the nucleus than the outer ones. Tables of dispersion correction to the scattering factors for various wavelengths are available[5,8,40] although, at present, these are mostly obtained from theoretical calculations, experimental measurements are scarce and not very reliable.

Although in the discussions in succeeding chapters, the correction terms are referred to as $\delta f'$ and $\Delta f''$, in the various formulas to be considered later, it becomes convenient to use (1) in the simplified form

$$f = f' + if'', \qquad (2)$$

where $f' = f_0 + \delta f'$ denotes the total real component of the scattering factor and f'' is used instead of $\Delta f''$, and denotes the imaginary component of f. We shall call $f' = f_0 + \delta f'$ the dispersion component of the scattering factor f, and f'' as the absorption component. These two components are responsible for the refractive index (deviation from unity), and the absorption coefficient of the medium, respectively.

The variation of $\delta f'$ and $\Delta f''$ as a function of v_i/v_K is shown in Fig. 1, where v_i is the frequency of incident radiation and v_K the K-absorption edge of the atom concerned. When v_i/v_K is larger than 1, both the components are finite, while for v_i/v_K smaller than 1, $\Delta f''$ is zero. The quantity $\delta f'$ is, in general, negative and hence it produces a decrease in the real component of the scattering factor of the atom from its normal value. The total real part of the scattering in the forward ($\theta = 0$) direction is not therefore equal to the total number of electrons in the atom. The case of $v_i/v_K > 1$ is of particular interest, since it leads to a phase change in scattering, this being due to $i\Delta f''$, which is always $\pi/2$ ahead of the real part. (The normal part, f_0, is actually behind the phase of the incident wave by π and hence $\Delta f''$ is equivalent to a lag of $\pi/2$ in phase relative to the incident wave. However, it is enough for our purposes to refer all phases with respect to that of the normal component f_0, since it is the same for all atoms.)

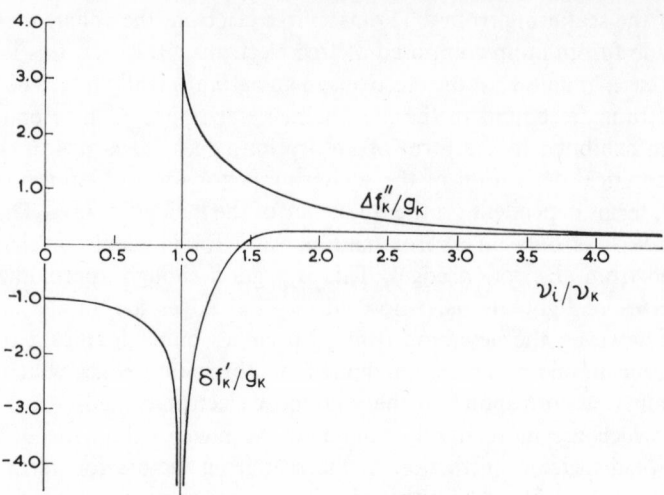

Fig. 1. Variation of $\delta f'$ and $\Delta f''$ with ν_i/ν_K for the K electrons. g_K is the oscillator strength for K electrons and is in general less than 2. (After James.[1])

The first and the most important consequence of this component with change in phase of scattering is the violation of Friedel's law[4] in noncentro-symmetric crystals; thus the intensities of x-rays scattered by a plane hkl is not equal to that by the inverse plane $\bar{h}\bar{k}\bar{l}$. Such an effect was first observed by Coster, Knol, and Prins,[4] who used the AuLα radiation ($\lambda = 1.2738$ Å) with ZnS, which has Zn atoms as anomalous scatterers, with their absorption edge at $\lambda_K = 1.2805$ Å. The anomalous component of S is negligible compared to that of Zn. The 111 and $\bar{1}\bar{1}\bar{1}$ reflections were shown to have different intensities. The deeper implications of this effect, from the point of view of structure analysis, were not fully appreciated until after nearly twenty years in 1949, when Bijvoet[1] and co-workers[2,27,28,30] showed that this method could be used for the unique determination of the absolute configuration of a molecule in a crystal, whose structure had been previously determined. It is also implicit in their results that the phase angle can be determined by this technique. The full potentialities of the method came to be realized soon and methods of applying them to actual structure determinations were evolved almost simultaneously, but independently, by different groups of workers.[19,20,27,29,35,37] The approach of Ramachandran and Raman[35,37] was basically from the point of view of Fourier syntheses and of the determination of the phases using the anomalous-dispersion effect. The approach of Pepinsky and co-workers[19,30,29] was from the point of view of the Patterson function in the presence of anomalous-dispersion effect and its deconvolution. Both approaches, in effect, lead to the structure in its absolute configuration.

Determination of the phase angle[27,35]

We consider the case of a noncentrosymmetric crystal containing N atoms, a part of which (say P atoms) are anomalous scatterers, while the rest, consisting of Q atoms ($Q = N - P$), are normal scatterers, whose anomalous components, if any, are negligible compared with those of the P atoms. In a number of practical examples, the anomalously scattering P atoms are all alike, and therefore we shall work out the formulas for this special case. We shall, however, point out the extension to the general case of nonidentical anomalous scatterers wherever necessary.

The relation between F'_P, F''_P, F_Q, and F_N are shown in the upper half of the Argand diagram in Fig. 2 for a reflection $\mathbf{H}(hkl)$. In this, $F_N(\mathbf{H})$ is simply

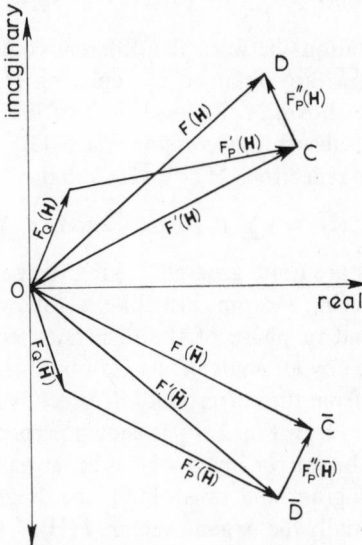

Fig. 2. Diagram showing the relationship between the various components of the structure factor for the inverse reflections $\mathbf{H}$ and $\mathbf{\bar{H}}$.

represented as $F(\mathbf{H})$ and the dispersion component of this is denoted[†] by $F'(\mathbf{H})$. We then have the relation

---

[†] The subscript N is optionally omitted in this chapter, where the emphasis is not on the number N.

$$F(\mathbf{H}) = F'(\mathbf{H}) + F''(\mathbf{H}),\tag{3}$$

where

$$F'(\mathbf{H}) = F'_P(\mathbf{H}) + F_Q(\mathbf{H}),\tag{4a}$$

and

$$F''(\mathbf{H}) = F''_P(\mathbf{H}).\tag{4b}$$

To be specific,

$$F'_P(\mathbf{H}) = \sum f'_{Pj} \exp(2\pi i \mathbf{H} \cdot \mathbf{r}_{Pj}),\tag{5a}$$

$$F''_P(\mathbf{H}) = i \sum f''_{Pj} \exp(2\pi i \mathbf{H} \cdot \mathbf{r}_{Pj}),\tag{5b}$$

and

$$F_Q(\mathbf{H}) = \sum f_{Qj} \exp(2\pi i \mathbf{H} \cdot \mathbf{r}_{Qj}).\tag{5c}$$

The corresponding relations between the different contributions to $F(\overline{\mathbf{H}})$ for the inverse reflection $\overline{h}\overline{k}\overline{l}$ are obtained by replacing $\mathbf{H}$ by $\overline{\mathbf{H}}(= -\mathbf{H})$ in (3), (4), and (5). These are shown in the lower half of Fig. 2. It will be noticed, in particular, that the absorption components $F''_P(\mathbf{H})$ and $F''_P(\overline{\mathbf{H}})$ contain a factor $+i$ for both the reflections $\mathbf{H}$ and $\overline{\mathbf{H}}$, so that

$$F''_P(\overline{\mathbf{H}}) = i \sum f''_{Pj} \exp(-2\pi i \mathbf{H} \cdot \mathbf{r}_{Pj}).\tag{5d}$$

The above formulas are quite general. Taking the particular case when all the anomalously scattering P atoms are alike, it follows that the absorption component F''_P is ahead in phase of the dispersion component F'_P by $\pi/2$, and is therefore rotated by an angle of $+\pi/2$ in both the top and the bottom halves of the diagram from the corresponding directions of F'_P. In the general case, this angle is not $\pi/2$. In Fig. 2, F''_P is shown perpendicular to F'_P.

We may now take the mirror image of the lower half of Fig. 2 in the real axis of the Argand diagram and thus obtain the diagram shown in Fig. 3, which represents essentially the Argand vectors $F(\mathbf{H})$, $F'(\mathbf{H})$, $\tilde{F}(\overline{\mathbf{H}})$, and $\tilde{F}''_P(\mathbf{H})$. Since $F'(\mathbf{H}) = \tilde{F}'(\overline{\mathbf{H}})$, this coincides with $F'(\mathbf{H})$ on reflection, and the triangle OCD of Fig. 2 becomes the triangle OCD' in Fig. 3. Since, in general, the vector CD is not perpendicular to OC, it is clear that the magnitudes of OD and OD' are not equal, in general, so that $|F(\overline{\mathbf{H}})|^2 \neq |F(\mathbf{H})|^2$. This is the phenomenon of the failure of Friedel's law, which occurs when a part of the atoms in the structure exhibit anomalous scattering of x-rays.

Denoting the angle between the vectors $F'(\mathbf{H})$ and $F''_P(\mathbf{H})$ of the reflection $\mathbf{H}$ by ψ, so that

$$\psi = \phi' - \phi''_P,\tag{6}$$

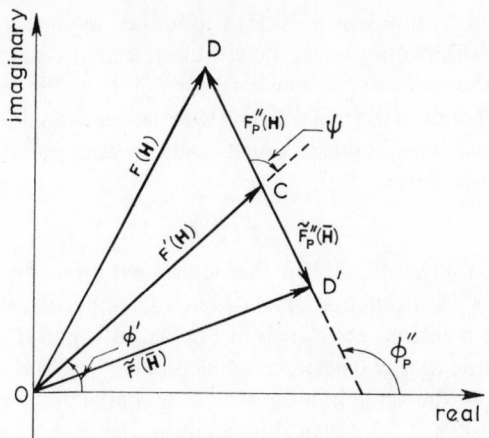

Fig. 3. Diagram showing the relation between $F(\mathbf{H})$, $\tilde{F}(\overline{\mathbf{H}})$ and $F_P''(\mathbf{H})$.

we have the following equations from Fig. 3.

$$|F(\mathbf{H})|^2 = |F'(\mathbf{H})|^2 + |F_P''(\mathbf{H})|^2 + 2|F'(\mathbf{H})|\,|F_P''(\mathbf{H})|\cos\psi, \qquad (7a)$$

$$|F(\overline{\mathbf{H}})|^2 = |F'(\mathbf{H})|^2 + |F_P''(\mathbf{H})|^2 - 2|F'(\mathbf{H})|\,|F_P''(\mathbf{H})|\cos\psi. \qquad (7b)$$

Adding the two, we obtain

$$|F'(\mathbf{H})|^2 + |F_P''(\mathbf{H})|^2 = \tfrac{1}{2}[|F(\mathbf{H})|^2 + |F(\overline{\mathbf{H}})|^2], \qquad (8)$$

while subtracting (7b) from (7a) gives

$$4|F'(\mathbf{H})|\,|F_P''(\mathbf{H})|\cos\psi = [|F(\mathbf{H})|^2 - |F(\overline{\mathbf{H}})|^2]. \qquad (9)$$

From (9) it follows that

$$\cos\psi = \frac{\Delta I}{4|F'(\mathbf{H})|\,|F_P''(\mathbf{H})|}, \qquad (10)$$

where ΔI stands for the difference in intensity of the pair of inverse reflections $\mathbf{H}$ and $\overline{\mathbf{H}}$, i.e.,

$$\Delta I(\mathbf{H}) = |F(\mathbf{H})|^2 - |F(\overline{\mathbf{H}})|^2. \qquad (11)$$

We shall call ΔI the *Bijvoet difference* for the reflection $\mathbf{H}$.

The results (8)–(11) are valid for the general case. The phase of $F'(\mathbf{H})$ is obviously given by

$$\phi' = \phi_P'' + \psi. \qquad (12)$$

Thus, if the intensities of the reflections $\mathbf{H}$ and $\overline{\mathbf{H}}$ are measured, and if the positions of the P atoms are known, then $F_P''(\mathbf{H})$ can be calculated, which yields $|F_P''(\mathbf{H})|$ and ϕ_P'' and hence, $F'(\mathbf{H})$ is obtained from (8). Thus, ψ can be obtained from (10), leading to the determination of the phase $\phi_N'(\mathbf{H})$ of the dispersion component of the structure factor $F_N'(\mathbf{H})$ of the reflection $\mathbf{H}$.

However, the determination of ψ from (10) is ambiguous, for only the cosine of the angle ψ can be obtained, which leads to two possible values, $\pm\psi$. Thus, (12) takes the form

$$\phi_N' = \phi_P'' \pm \psi, \tag{13}$$

and there is a 2-fold ambiguity in the determination of the phase, exactly as in the case of isomorphous crystals (in the noncentrosymmetric case). The two possible solutions are shown in Fig. 4a, in which $|F(\mathbf{H})|$, $|\tilde{F}(\overline{\mathbf{H}})|$ and $F_P''(\mathbf{H})$ are the same, but ψ has two possible values $+\psi$ and $-\psi$.

Figure 4b shows the same in a construction similar to Fig. 1 of Ch. 9 for isomorphous crystals. In drawing this diagram, the direction of the Argand vector F_P'' is taken to be perpendicular to F_P, which, as already stated, is true if all the anomalously scattering atoms are alike. The two solutions, which occur on the right- and left-hand sides of this figure, are obtained by reflection of each other about the line MOM' parallel to F_P', passing through the origin. Thus, in this special case, the two possible values for ϕ_N' are

$$\phi_N' = \phi_P' + \frac{\pi}{2} \pm \psi. \tag{14}$$

If the anomalously scattering P atoms form a set of replaceable atoms in the structure, then, form the results obtained in Ch. 9, the ambiguity in phase, as determined by the isomorphous replacement method, would be of the form

$$\phi_N' = \phi_P' \pm \theta \tag{15}$$

[cf. Fig. 1 and (4) of Ch. 9]. This corresponds to the two possible directions of F_N' being related by reflection about the line NON' parallel to F_P' through the origin. Since, in general, the two vectors F_P' and F_P'' do not coincide (in the special case of identical atoms they are in fact perpendicular), the second solution introduced by the experimental method is not the same for the isomorphous crystal technique and the anomalous dispersion technique. This fact can be analytically expressed in the equal-atom case by saying that, if θ $(= \phi_N' - \phi_P')$ is the correct solution, the two ambigous values in the isomorphous crystal method are $+\theta$ and $-\theta$, while, in the anomalous-dispersion method, they are θ and $\pi - \theta$, since $\psi = \pi/2 - \theta$ in this case.

Resolution of phase ambiguity

In most practical cases, it turns out that the anomalous scatterers are heavy atoms. This is a fortunate circumstance for, in many cases of isomorphism met with in practice, the replaceable atoms are also heavy atoms like Br, Cl,

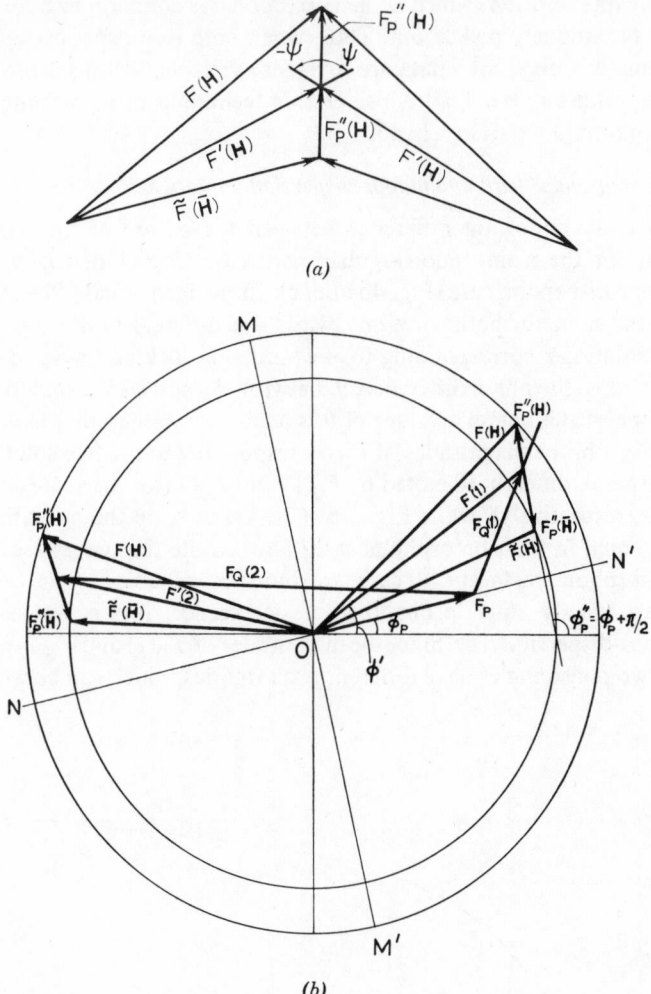

Fig. 4. (*a*) Two possible solutions for the angle ψ. (*b*) The two solutions drawn in a diagram analogous to the isomorphous case, Figure 1 of Ch. 9.

or heavy metal ions. Thus it is possible to find a situation in which we have x-ray intensity data for a pair of isomorphous crystals, and in addition, anomalous-dispersion data for one of the compounds. Restricting ourselves to the case when all the P atoms are alike, we obtain the result that the isomorphous-replacement data gives for ϕ an ambiguity $\phi'_P \pm \theta$, while the anomalous-dispersion data gives the pair of values $\phi'_P + \theta$ and $\phi'_P + \pi - \theta$.

Thus only one solution (which is the correct one) is common to both sets, and this can be uniquely picked out. Thus, when both isomorphous-replacement and anomalous-dispersion data are together available, the phase problem has a unique solution. We shall consider this technique of combining the two methods more in detail in Ch. 13.

Resolution of phase with anomalous-dispersion data alone

There is an important difference between the nature of the phase circle diagrams for the isomorphous-replacement case (Fig. 1 of Ch. 9) and the anomalous-dispersion case (Fig. 4b above). In the former case, the magnitude of F_Q is the same for both solutions, while it is different in the latter case for the two solutions, corresponding to $+\psi$ and $-\psi$. Taking the special case of equal P atoms, the phase difference, θ, between ϕ_N' and ϕ_P' is equal to $\pi/2 \pm \psi$, so that one of the possible values of θ is acute, while the other is obtuse. As seen in Fig. 4b, the magnitudes of F_Q corresponding to the two solutions of ψ are, in general, different (denoted by $F_Q(1)$ and $F_Q(2)$ for the acute and obtuse solutions, respectively), while Fig. 1 of Ch. 9 shows, on the other hand, that they are equal in the isomorphous case. Thus, while the two phase angles in the isomorphous-replacement case are indistinguishable, in the sense that they have exactly equal probability of occurrence, this is not so for the anomalous-dispersion case. Since the magnitudes of $F_Q(1)$ and $F_Q(2)$ are different, the two phase angles have different probabilities, which can be worked out

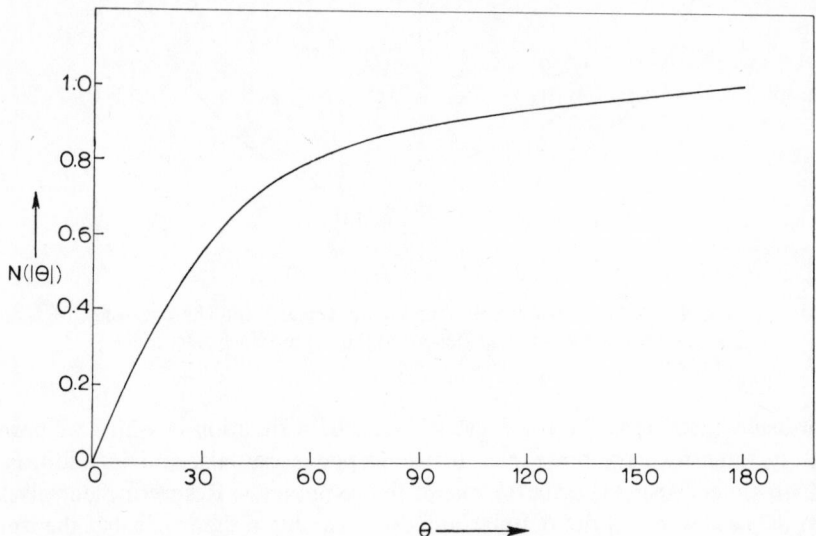

Fig. 5. Variation of $N(|\theta|)$ with $|\theta|$ for $\sigma_1^2 = 0.6$. The data correspond to the mean of cases (B) and (C) of Table 1.

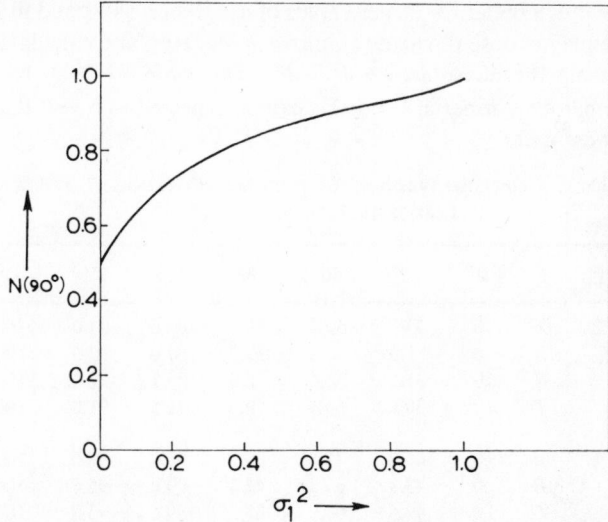

Fig. 6. Variation of $N(90°)$, the probability that $\theta, = \phi_N - \phi'_P$, be acute, with σ_1^2. The curve corresponds to the means of cases (B) and (C) of Table 1.

from theory, since the relevant probability distribution is known,[36] and the probability falls down with increase in magnitude of $|F_Q|$. Using this result, the probability distribution of the phase angle difference, θ, can be determined[24] and hence the probability that θ be acute, which may be denoted by $N(90°)$, can be calculated.[24] This is found to be a function of σ_1^2, the fractional mean contribution to the intensity from the P atoms. It also depends on the nature of the P group of atoms, namely whether it consists of one, two, or many atoms, and, in the last case, whether the group has a center of symmetry or not. The results for $N(\theta)$ for all these cases are given in Table 1. The mean of the values for the two-atom P group and one consisting of many atoms in a noncentrosymmetric configuration (which are very close to each other) is shown diagrammatically in Figs. 5 and 6, for a typical variation of $N(|\theta|)$ with $|\theta|$ and for the variation of $N(90°)$ with σ_1^2, respectively.

Quasianomalous-dispersion synthesis

It will be seen from Table 1 and Fig. 6 that, for σ_1^2 as low as 20 per cent, nearly 75 per cent of the reflections have an actue value for θ ($= \phi'_N - \phi'_P$). Thus, of the two possible solutions provided by the anomalous-dispersion method, Fig. 4b, if we choose the acute value $\theta(I)$, then the probability that it is correct is three times that it is wrong, even for $\sigma_1^2 \simeq 20$ per cent and the chances are much better for larger values of σ_1^2. Hence, as a good first approximation, we may choose the acute solution in all cases, and calculate a Fourier synthesis using the phase ϕ_N^{l} ($= \phi_P' + \theta^{\mathrm{l}}$). This synthesis may be called the "quasianomalous synthesis," and this can be expected to reveal the unknown atoms in most cases.

Table 1   Cumulative function $N(|\theta|)^a$ for the different cases^b at different heavy-atom contributions σ_1^2 ^{24,34}

σ_1^2	$|\theta|$	0°	30°	60°	90°	120°	150°	180°
0.2	A	0	33.7	59.4	76.0	86.6	94.0	100
	B	0	31.5	56.6	73.0	84.0	92.5	100
	C	0	31.2	56.0	72.0	83.7	92.0	100
	D	0	30.0	53.5	69.5	81.5	91.0	100
0.4	A	0	46.2	75.5	88.5	94.4	98.0	100
	B	0	43.3	69.2	82.3	90.0	95.5	100
	C	0	42.5	68.0	81.5	89.5	95.0	100
	D	0	39.5	63.5	77.5	86.5	93.2	100
0.6	A	0	62.0	88.5	96.0	98.1	99.0	100
	B	0	56.5	80.1	89.0	94.0	97.6	100
	C	0	55.0	78.9	88.5	93.8	97.2	100
	D	0	50.0	72.3	83.0	90.2	95.0	100
0.8	A	0	83.0	98.7	99.8	99.9	99.9	100
	B	0	72.5	88.6	93.5	96.3	98.0	100
	C	0	71.5	89.3	94.5	97.1	98.0	100
	D	0	63.2	81.6	89.0	93.7	97.0	100

^{a} $N(|\theta|)$ is the fraction of reflections which have $|\phi - \phi_P| < |\theta|$ and is given in per cent.
^{b} The cases are those in which the known P group consists of (A), one atom; (B), two atoms; (C), many atoms, noncentrosymmetric; (D), many atoms, centrosymmetric.

The basic idea of the quasianomalous-dispersion synthesis, namely that most of the reflections have an acute value of θ ($= \phi_N' - \phi_P'$), was derived from general considerations in the authors' laboratory[35] as early as 1956, long before the detailed theory described above was worked out (see also Ref. 32). This deduction was also verified experimentally[35] in the case of the $hk0$ reflections of ephedrine hydrochloride.[31] The results of this study are given in Table 2. The measured anomalous dispersion data are tabulated in the form of the Bijvoet ratios [r, defined by (16) below], rather than as Bijvoet differences, ΔI.

$$r = \frac{I(\mathrm{H}) - I(\overline{\mathrm{H}})}{\frac{1}{2}[I(\mathrm{H}) + I(\overline{\mathrm{H}})]} = \frac{\Delta I}{I_{\mathrm{mean}}}. \tag{16}$$

This is because the evaluation of ΔI requires the absolute scale factor to be known, while the Bijvoet ratio can be calculated straightaway from the observed data. The value of ϕ'_P, and the two values of ϕ'_N deduced from the anomalous dispersion data, are tabulated in column 4 and in columns 5 and 6, respectively. Of these, the one which is closer to the heavy atom phase, ϕ'_P, is shown as in $\phi'^{(I)}_N$. In the majority of cases, this is close to the true phase, calculated from the known structure. The cases where they disagree are mostly weak reflections, which at the same time have exceptionally large magnitudes for the Bijvoet ratios. This is because $|F'_N|$ and $|F''_P|$ are of nearly the same order of magnitude for these examples.

It is thus seen that most of the medium and strong reflections are correctly phased by the method of choosing the phase closer to that of the heavy anomalous scatterer. Leaving aside the examples of the weak reflections, in which there is complete disagreement, the mean error in phase, determined by the anomalous-dispersion technique, is only 12° for the first 20 reflections, and 18° for the remaining weaker reflections. Thus, this early experimental study clearly showed that phases could be determined to a good degree of accuracy in a noncentrosymmetric crystal purely from measurements of Bijvoet differences. In fact, following the success of the experiments in ephedrine hydrochloride, Raman[38] solved the crystal structure of lysine hydrochloride dihydrate, using the anomalous-dispersion effects of the chlorine atom. We shall consider this, and other applications of anomalous dispersion for structure analysis, via the quasianomalous method, in a later section of this chapter.

Application of the quasianomalous-dispersion synthesis

We shall now discuss the considerations which are involved in the application of the quasianomalous-dispersion synthesis, for which we shall broadly follow the argument in a recent short review.[34] As mentioned above, the fractional number of acute values of θ, namely $N(90°)$, is a function of σ_1^2, increasing with increasing value of σ_1^2, the heavy-atom contribution. The quasianomalous-dispersion synthesis was therefore calculated in a hypothetical case, having $\sigma_1^2 = 10$ per cent. The example in which this was tested had 24 carbon and 22 oxygen atoms in the unit cell—in the same locations as in the structure of cellobiose,[14] with two additional anomalously scattering atoms, each with a scattering power of half that of a chlorine atom. The Fourier synthesis, obtained by using the phase closer to that of the heavy atom, as determined by (10), is shown in Fig. 7. It will be noticed that this diagram contains clear peaks at almost all the unknown atomic locations, and has very few false peaks.

Table 2 Measurement of anomalous dispersion effect[35] for the $hk0$ reflections of ephedrine hydrochloride[a]

h	k	l	Structure amplitude $\|F\|$	Bijvoet ratio $\Delta I/I_{mean}$ %	Phase of anomalous scatter ϕ'_P	Two solutions of phase[b] $\phi'_N(1)$	$\phi'_N(2)$	Phase from structure ϕ'_N	Validity of hypothesis[c]
1	1	0	18.5	−11.0	180	152	28	61	×
2			25.2	−10.4	180	147	33	158	✓
3			40.0	−2.4	180	166	14	174	✓
4[a]			16.8	−2.0	—	—	—	—	—
5			14.4	20.6	0	53	127	113	×
6			27.5	0.0	0	0	180	11	✓
7			7.4	1.0	0	1	179	4	✓
8			8.0	0.0	—	—	—	—	—
9			9.4	12.8	180	198	342	225	✓
10			8.8	−9.2	180	172	8	168	✓
11			5.5	0.0	180	180	0	175	✓
12			2.7	−3.2	—	—	—	—	—
0	2	0	17.8	0.0	180	180	0	174	✓
1			20.9	−5.0	180	164	16	153	✓
2			12.9	0.0	—	—	—	—	—
3			4.9	54.0	0	58	122	116	×
4			19.5	−13.6	0	332	208	324	✓
5			7.0	−18.8	0	339	201	325	✓
6			0.0	0.0	—	—	—	—	—
7			8.1	−13.4	180	164	16	163	✓
8			11.5	0.0	180	180	0	172	✓
9			12.3	4.6	180	189	351	220	✓
10			7.2	−7.6	—	—	—	—	—
11			6.4	−7.0	0	355	185	351	✓
12			7.8	14.2	0	11	169	27	✓

h	k	l							
1	3	0	15.8	-4.8	0	347	193	349	√
2			17.2	1.8	0	3	177	15	√
3			14.8	5.2	0	11	169	10	√
4			12.5	-5.0	—	—	—	—	—
5			8.8	7.0	180	191	349	194	√
6			6.8	27.0	180	202	338	242	√
7			3.5	-12.0	180	168	12	140	√
8			10.7	-8.0	—	—	—	—	×
9			3.9	-5.2	0	358	182	146	√
10			9.6	-11.6	0	349	191	329	√
11			3.7	9.2	0	5	175	18	√
0	4	0	12.7	9.2	0	10	170	22	√
1			9.0	1.4	0	2	178	7	√
2			5.3	-2.4	—	—	—	—	×
3			3.3	43.6	180	195	345	280	×
4			10.1	8.6	180	188	352	179	√
5			8.6	4.6	180	188	352	202	√
6			1.9	—	—	—	—	—	—
7			4.7	-54.0	0	302	238	273	√
8			9.4	4.8	0	5	175	26	√
1	5	0	7.6	-27.2	180	150	30	124	√
2			12.1	-11.0	180	163	17	179	√
3			4.7	-24.4	180	161	19	153	√
4			2.0	-13.2	—	—	—	—	—
0	6	0	8.6	-5.0	180	175	5	161	√

[a] Reflections for which columns beyond the fourth are left vacant are those for which θ is indeterminate, since $|F_P''| \simeq 0$.

[b] The one closer to the heavy-atom phase is shown as $\phi_N'(1)$.

[c] Hypothesis is that the phase $\phi_N'(1)$ is close to the true phase ϕ_N'.

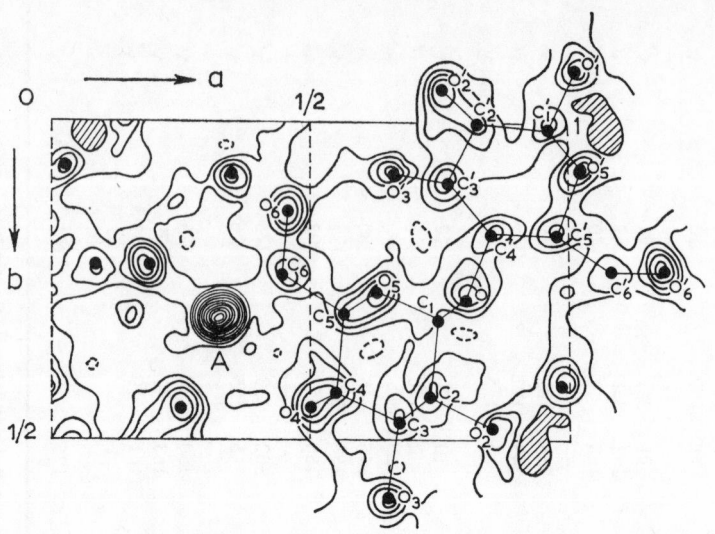

Fig. 7. The quasianomalous synthesis for $\sigma_1^2 = 10$ per cent.[14] The contour interval is $2e\text{A}^{-2}$ and the first contour is at $2e\text{A}^{-2}$. The unknown atoms are represented by dots and the known anomalous scatterers by the symbol A.

Magnitude of Bijvoet ratios

The above test, however, gives an over-optimistic picture, for the anomalous-dispersion data have been assumed to have no errors in them. In order to assess the possible accuracy of the data, it would be worthwhile to know the mean value of the Bijvoet ratio, r. The mean value $\langle r \rangle$ has been calculated from theory[23] for the four types of composition of the anomalous scatterers listed in Table 1, and the values thus obtained are given in Table 3 for different values of σ_1^2. It will be seen from this table that the Bijvoet ratio is largest for a structure with $\sigma_1^2 \simeq 0.5$, and that it decreases both for smaller, as well as larger, values of σ_1^2. The data given in Table 3 are for a value of $f_P''/f_P'(=k)$ equal to 0.2, which is approximately the order of this quantity for several anomalous scatterers for Cu$K\alpha$ and Mo$K\alpha$ radiations. The mean value of k for the range of the Bragg angle from 0 to 50° are given in Table 4 for several medium and heavy atoms. The magnitudes of $\langle r \rangle$ for values of k varying from 0.04 to 0.30 have also been tabulated[23] and this table will be valuable for investigators planning the use of the anomalous dispersion method.

Although the mean value of the Bijvoet ratio, $\langle r \rangle$ is largest for σ_1^2 equal to about 0.45 (for $k = 0.2$) and is of the order of 40 per cent, the value of $\langle r \rangle$ is of the order of 25 per cent even for σ_1^2 as low as 0.1, for $k = 0.2$. Hence, even for

Table 3 Mean value of the Bijvoet ratio r as a function of σ_1^2

σ_1^2	Structure type for P atom[a]			
	A	B	C	D
0.1	0.2796	0.2460	0.2400	0.2098
0.2	0.3831	0.3289	0.3200	0.2738
0.3	0.4478	0.3750	0.3666	0.3094
0.4	0.4824	0.3951	0.3919	0.3290
0.5	0.4857	0.3928	0.4000	0.3370
0.6	0.4529	0.3704	0.3919	0.3354
0.7	0.3795	0.3315	0.3666	0.3237
0.8	0.2720	0.2800	0.3200	0.2993
0.9	0.1609	0.2129	0.2400	0.2514

[a] The cases are those in which the known P group consists of, (A), one atom; (B), two atoms; (C), many atoms, noncentrosymmetric; and (D), many atoms, centrosymmetric.

a crystal in which the anomalous scatterers contribute as low as 10 per cent to the mean intensity, theory predicts that the difference between $I(\mathbf{H})$ and $I(\overline{\mathbf{H}})$ is readily measurable for a majority of reflections.

Table 4 Mean value of $k(=f''/f')$ for some typical atoms which scatter $CuK\alpha$ and $MoK\alpha$ radiations anomalously

Atom	Atomic number	Value of k for	
		$CuK\alpha$	$MoK\alpha$
S	16	0.07	0.04
Cl	17	0.08	0.04
Ca	20	0.14	0.05
Cr	24	0.21	0.09
Mn	25	0.23	0.10
Fe	26	0.25	0.11
Co	27	0.30	0.12
Zn	30	0.04	0.15
Br	35	0.07	0.20
I	53	0.22	0.10
Pt	78	0.16	0.28
Hg	80	0.17	0.30

Crystal structures solved by anomalous-dispersion method

The data shown in Table 2 indicate clearly that it is possible to determine the phases of a majority of reflections by choosing out of the two ambiguous solutions obtained by measuring the Bijvoet differences, the one closer to that of the anomalous scatterers. In view of this result, Raman[38] attempted to solve the structure of a noncentrosymmetric crystal, that of L-lysine hydrochloride dihydrate, by applying this technique. The relevant values of σ_1^2 and k for this structure were 0.49 and 0.08, as compared with $\sigma_1^2 = 0.44$ and $k = 0.08$ for the case of the test substance, ephedrine hydrochloride. The method was successful, and purely by making use of measurements of anomalous-dispersion data in two noncentrosymmetric zones of reflections, and calculating the quasianomalous-dispersion synthesis in each case, the atomic parameters could be determined.[38] The Fourier maps obtained using the phases deduced from anomalous-dispersion data are shown in Figs. 8a and 8b.

Since then, a number of crystal structures have been solved using the quasianomalous-dispersion synthesis; a list of these is given in Table 5, which contains also the values of σ_1^2 for these cases. It will be seen that σ_1^2 ranges from 0.26 to 0.93. The theoretically expected mean values of the Bijvoet ratio, as well as the actually observed values, where available, are also given in Table 5. This table shows that the Bijvoet ratio is in the neighborhood of 20

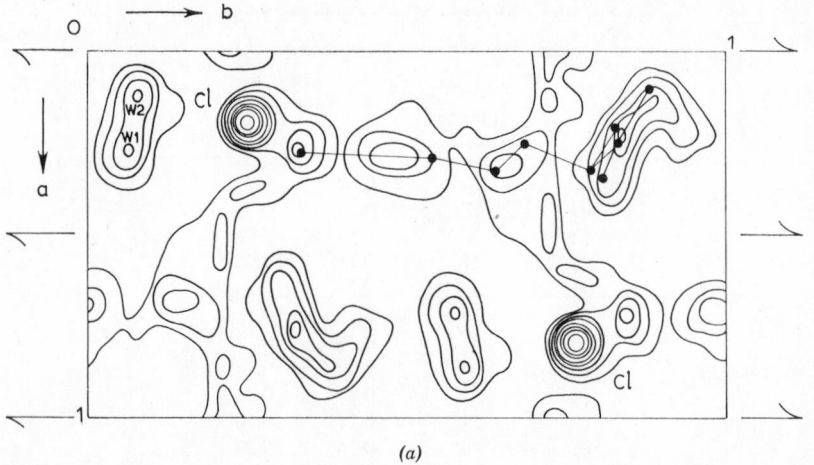

(a)

Fig. 8. The anomalous dispersion Fourier synthesis obtained with lysine hydrochloride dihydrate[18] *(a)* projected on (001);

Table 5 Structures solved by using the quasianomalous synthesis

Crystal	Space Group	Anomalous scatterer	$\sigma_1^{2\,a}$ %	k^a %	$\langle|\Delta I|/I\rangle$ Theory[b]	Theory[c]	obs[d]	Reference
L-Ephedrine HCl[e] ($C_{10}NOH_{1.5}\cdot HCl$)	$P2_1$	Cl	44	8	15	11	11	Ramachandran and Raman[35]
L-Lysine HCl·$2H_2O$ ($C_6N_2O_2H_{14}HCl\cdot 2H_2O$)	$P2_1$	Cl	49	8	15	11	11	Raman[38]
Factor V 1a ($C_{46}H_{66}O_9N_{11}Co.\ 11\ H_2O$)	$P2_1$	Co	26	30	50	36	~20	Dale, Hodgkin and Vankatesan[7]
Cytisine HBr·H_2O ($C_{11}H_{14}N_2O_2Br$)	$P2_1$	Br	85	20	25	21	13	Geurtz[9]
Cytisine HI·H_2O ($C_{11}H_{14}N_2O_2I$)	$P2_1$	I	93	22	20	18	—	Geurtz[9]
Methyl melaleucate iodoacetate ($C_{34}H_{51}O_5I$)	$P2_12_12_1$	I	83	22	30	25	—	Hall and Maslen[10]
Davallol iodoacetate ($C_{32}H_{51}O_2I$)	$P2_12_12_1$	I	87	22	25	24	—	Oh and Maslen[18]
α-Quartz[e] (SiO_2)	$P3_12$	Si	84	5	7	6	7	Zachariasen[44]

[a] Values of σ_1^2 and k are the average values over the range $\theta \leqslant 50°$ for CuKα radiation, and MoKα in the case of cytisine HBr·H_2O.

[b] Theory of Parthasarathy.[23]

[c] Theory of Zachariasen[44] (approximate).

[d] The values in this column are the mean for the measured reflections. Weak reflections were not measured, in general, except in the case of quartz.

[e] Structures for which anomalous-dispersion data are available, but whose crystal structures were already known. Ephedrine was the first structure for which phase determination by the anomalous-dispersion method was tested. The most accurately measured data for anomalous dispersion effects are for quartz.

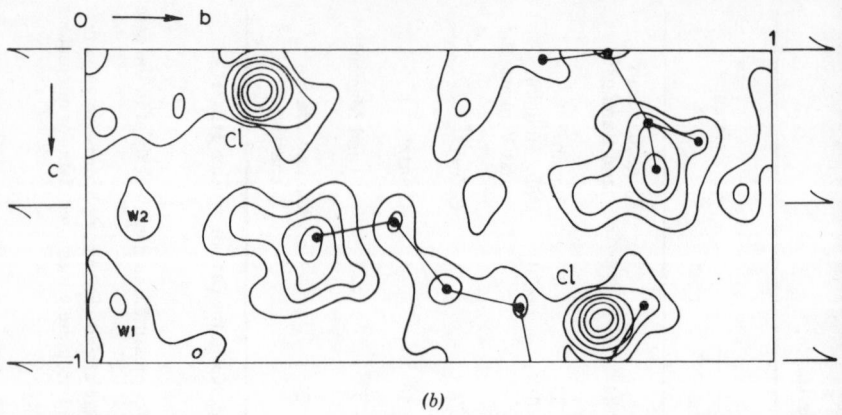

(b)

Fig. 8. (*b*) projected on (100). The atomic positions are marked by dots and one molecule is shown joined.

per cent in many of the crystal structures that have been investigated. The observed values (for the medium and strong reflections) have magnitudes which are lower than that given by theory. Where all reflections have been measured, as in quartz, the agreement between theory and experiment is good. It is interesting that the very first example[38] had a value of k as low as 0.08, with also a medium value of $\sigma_1^2 = 0.49$ and a mean $\langle r \rangle$ of 10–15 per cent. The lowest value of σ_1^2 for which a successful application of the quasianomalous-dispersion method has been made is 0.26, in the case of factor VIa. This has a cobalt atom as anomalous scatterer in a molecule containing 66 light (C, N, O) atoms.

Power of the anomalous-dispersion method

Theory shows that, using cobalt as anomalous scatterer for CuKα radiation, a structure having as many as 200 light atoms (nitrogens) is amenable for study with this method ($\sigma_1^2 = 0.1$, $\langle r \rangle \simeq 40$ per cent), whereas by using iodine as anomalous scatterer, a molecule with as many as 500 light atoms can be solved ($\sigma_1^2 = 0.2$, $\langle r \rangle \simeq 35$ per cent).[†] A doubt may arise, in this connection, as to the reasonableness of taking the *mean* value $\langle r \rangle$ as an estimate of the measurability of the Bijvoet ratio. Thus, it may be questioned whether a large mean value is produced by a small number of weak reflections having very large Bijvoet ratios, while the majority of them have small values of r. This question can be settled only by calculating the *distribution* of r, which un-

---

[†] We are grateful to Dr. S. Parthasarathy for the estimates in this paragraph, which were calculated by him from his theory.[23]

fortunately has not been possible to work out theoretically so far. However, the cumulative function $N(r)$ (which gives the probability that the Bijvoet ratio is less than r) has been calculated for a practical case (hypothetical structure), with $\sigma_1^2 = 0.5$ and $k = 0.2$.[34] This is shown in Fig. 9, from which

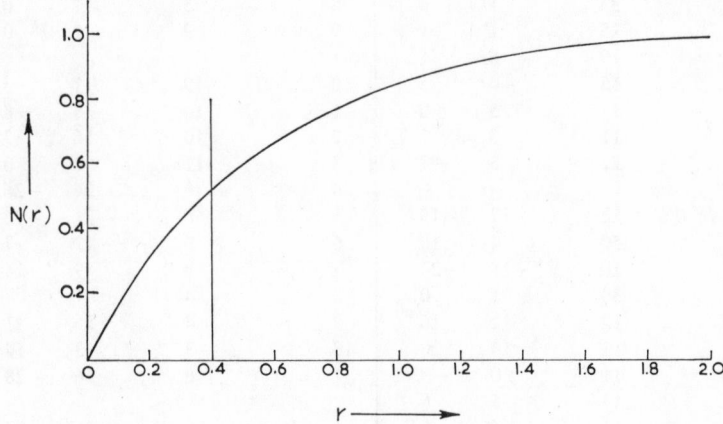

Fig. 9. Cumulative function $N(r)$ of the Bijvoet ratio r, for $\sigma_1^2 = 0.5$ and $k = 0.2$.[21] The mean value is marked by a vertical line.

it can be seen that, while the mean value $\langle r \rangle$ is 40 per cent, more than 80 per cent of the reflections have a measurable Bijvoet ratio of greater than 10 per cent. It is reasonable to expect that large Bijvoet ratios will, in general, be associated with weak reflections, since this requires that $|F_N'|$ and $|F_P''|$ should be of the same order of magnitude. This explains why the measured mean Bijvoet ratios, listed in the last column of Table 5, are less than the calculated mean values, for the anomalous-dispersion effects are difficult to measure accurately for weak reflections. An idea of the actual distribution of the observed Bijvoet ratios may be obtained from Tables 6 and 7, in which the observations made on three crystals by Geiger-counter methods are summarized. Although counter methods were used in the earliest studies[35,38] made in Madras, and in one later investigation,[9] only photographic methods have been applied in the more recent investigations.

The above discussion indicates the great importance of making accurate intensity measurements of pairs of inverse reflections and the need for having diffractometer arrangements in which this can be done at a single setting of the crystal (as, for example, by rotating through an angle of 180° about the ω circle with a four-circle diffractometer). If the relative intensities of inverse reflections can be measured to an accuracy of better than 5 per cent (an

Table 6 Bijvoet-ratio data for L-tyrosine HCl[22]

h	k	l	$|F_{obs}|$	Exp. %	Calc. %	h	k	l	$|F_{obs}|$	Exp. %	Calc. %
1	1	0	18	2	4	7			9	7	10
3			33	1	4	8			5	7	0
4			25	1	1	9			9	−1	0
5			9	8	11						
6			15	−4	−5	0	6	0	19	−1	−3
7			17	5	0	1			14	−5	−8
8			11	0	−7	2			10	16	13
9			22	−5	−7	3			11	2	0
						4			8	−24	−26
0	2	0	22	−17	−19	5			6	−21	−42
1			50	4	10	6			9	−12	7
2			10	−18	−24	7			8	9	14
3			30	1	0	8			4	−1	7
4			12	12	−16	9			4	25	11
5			11	14	20	10			3	13	14
6			11	0	−9	11			6	−36	−28
7			17	5	6						
8			11	−6	−8	1	7	0	6	21	44
9			14	2	0	2			7	24	14
11			10	9	−7	3			10	21	15
12			9	0	0	4			9	1	0
						5			6	−7	−12
1	3	0	15	−13	−19	7			7	−2	1
2			31	0	−4	8			11	−8	−15
3			26	2	0	10			6	−10	−18
4			22	6	−4						
5			7	18	15	0	8	0	8	11	13
6			9	0	−13	1			11	−7	−16
8			11	0	−14	2			7	−23	−24
9			6	34	32	3			9	1	0
10			9	3	8	4			6	−4	−15
						5			6	−14	−12
0	4	0	30	9	10	6			6	35	20
1			17	19	15						
2			19	2	−8	1	9	0	3	−59	−31
3			11	1	0	2			4	20	26
4			4	41	23	3			8	3	0
5			8	19	−8	4			7	11	0
6			15	3	0	5			4	4	−10
7			15	−4	−7	6			5	13	6
10			10	9	−7						
11			8	3	−5	1	10	0	6	19	20
						4			4	−1	0
1	5	0	11	−13	−16	5			4	−12	−10
2			9	−17	−15						
3			16	1	0	1	11	0	3	−8	−34
4			12	−2	0	2			4	−36	−46

Table 7   Bijvoet-ratio data for cytisine HBr·H_2O [9]

h	k	l	$\|F_{obs}\|$	Exp. %	Calc. %	k	k	l	$\|F_{obs}\|$	Exp. %	Calc. %
0	1	1	6	87	184			3	29	−1	0
		2	62	−13	−15			4	15	4	6
		3	40	−9	−6			5	—	—	—
		4	14	40	38			6	12	−19	−7
		5	12	3	4						
		6	20	−2	0	0	6	0	40	−9	−2
		7	27	1	0			1	22	−19	−11
								2	6	—	—
0	2	0	64	13	5			3	21	26	21
		1	25	15	14			4	26	−6	7
		2	38	3	4			5	23	−8	−10
		3	6	38	30						
		4	23	9	4	0	7	1	23	6	−3
		5	25	10	7			2	27	−8	−7
		6	17	−6	−7			3	10	−18	−12
								4	21	13	20
0	3	1	48	10	11			5	9	11	0
		2	52	−1	0						
		3	50	−4	−4						
		4	7	—	—	0	8	0	19	−5	−9
		5	8	31	26			1	14	−20	−20
		6	17	—	—			2	10	−6	−4
								3	9	34	30
0	4	0	39	−34	−33			4	10	−5	8
		1	42	−11	−8						
		2	22	13	13	0	9	1	9	7	24
		3	27	4	0			2	16	9	10
		4	26	−5	−3			3	12	−10	−7
		5	17	−8	−5						
0	5	1	28	3	1	0	10	0	6	—	—
		2	24	23	21			1	17	−3	−2

overall accuracy which has been achieved in quite a few investigations in recent years) the quasianomalous-dispersion method is likely to be extremely valuable in solving crystal structures which have a suitably chosen single set of heavy atoms.

The quasianomalous-dispersion method has been applied for phase determination in some proteins. However, in all these cases, the anomalous dispersion method was not used alone, but in combination with the isomorphous-addition method.

The weighted anomalous-dispersion synthesis

In the quasianomalous-dispersion synthesis, which we have been discussing so far, one of the two possible ambiguous solutions for the phase (namely the one which is closer to that of the heavy atom contribution) is used for the synthesis. This choice has always a higher probability of being correct than the other one. However, instead of always choosing the value which has the higher probability, we could assign relative probabilities to the two possible values and include both the solutions for the phase angle in the synthesis. As indicated below, this can be shown to be equivalent to multiplying the structure factor by a suitable weighting function. The synthesis obtained by using such weighted structure factors may be called the "weighted anomalous-disperion synthesis." We shall briefly consider its theory and properties.

Since $F'_N = F'_P + F_Q$, it follows that $|F_Q|^2 = |F'_N - F'_P|^2$. Hence, we may substitute this in the expression for $P(F_Q)$, namely,

$$\dot{P}(F_Q) = \frac{1}{\pi S_Q^2} \exp\left(\frac{-|F_Q|^2}{S_Q^2}\right),\tag{17}$$

and obtain

$$P(F_Q) = \frac{1}{\pi S_Q^2} \exp\left[\frac{-|F'_N - F'_P|^2}{S_Q^2}\right],$$

$$= \frac{1}{\pi S_Q^2} \exp\left[\frac{-(|F'_N|^2 + |F'_P|^2)}{S_Q^2}\right] \exp\left(-X\cos\theta\right),\tag{18}$$

where

$$X = \frac{2|F'_N|\,|F'_P|}{S_Q^2}\tag{19}$$

and

$$\sin\theta = \frac{\Delta I}{4|F'_N|\,|F''_P|},\tag{20}$$

which follows from (10). Denote the acute solution for θ of (20) by θ_0. The two ambiguous solutions are then equal to θ_0 and $\pi - \theta_0$. If these values are substituted in (18), and the weighting factor W_A is worked out, it is found to have the form[25]

$$W_A = \cos\theta_0 \tanh\left(X\cos\theta_0\right) + i\sin\theta_0,\tag{21}$$

which can be put in the form

$$W_A = |W_A| \exp (i\phi_W), \qquad (22)$$

where

$$|W_A| = \{[\cos \theta_0 \tanh (X \cos \theta_0)]^2 + \sin {}^2\theta_0\}^{1/2}, \qquad (23a)$$

and

$$\phi_W = \tan^{-1} \left[\frac{\tan \theta_0}{\tanh (X \cos \theta_0)} \right]. \qquad (23b)$$

Thus, the coefficients to be used in the weighted anomalous-dispersion synthesis are $|W_A| |F_N| \exp i(\phi_P' + \phi_W)$, where $|W_A|$ and ϕ_W are given by (23a) and (23b). It is readily seen that the weighting function can be calculated from the known positions of the anomalous scatterers, the known contents of the unit cell and the measured Bijvoet differences, and, of course, the mean structure amplitudes. Essentially similar ideas regarding the possible weighting of anomalous dispersion data have been given in a note by Sim.[41]

Since the best available statistical information has been used in it, the weighted anomalous-dispersion synthesis, $W_A F_N \exp (i\phi_P)$, is expected to be superior to the simple quasianomalous-dispersion synthesis in revealing the unknown Q atoms. A test of this synthesis, in the same example that was used

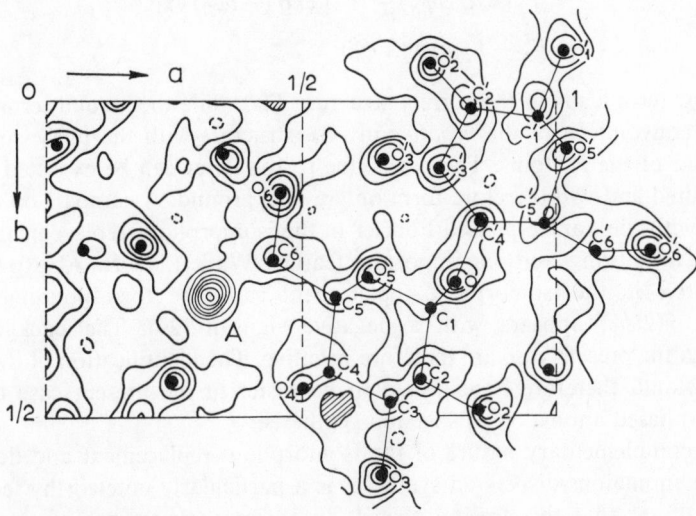

Fig. 10. The weighted-anomalous synthesis of the same structure as considered in Figure 7. All details are as in Figure 7. Note the relative decrease in the peak height of the known atom, while the peak heights at the unknown atoms are nearly the same as in Figure 7.

for obtaining the quasianomalous-dispersion synthesis in Fig. 7, is shown in Fig. 10. It is found that most of the Q atoms have come out stronger in the weighted anomalous-dispersion synthesis (Fig. 10) than in the quasi-anomalous-dispersion synthesis (Fig. 7), *relative* to the P atoms. Actual measurement shows that the ratio of the peak strengths at the unknown Q atoms to those at the known P atoms has improved by a factor of 1.8, on the average. However, in actual structure analysis of unknown crystal structures, the weighted anomalous-dispersion synthesis has not been found to be particularly superior to the single-phased, quasianomalous-dispersion synthesis.

The double-phased synthesis[32,33]

We shall now consider the properties of a Fourier synthesis obtained by using both the phases given by (10). We may call this the double-phased anomalous-dispersion Fourier synthesis.

The two possible phase angles given by (10) can be written more explicitly as ϕ'_N and $(2\phi'_P + \pi - \phi'_N)$. When both these phases are used in a Fourier synthesis, with $|F'_N|$ as amplitudes, we obtain the coefficients

$$|F'_N| \exp(i\phi'_N) + |F'_N| \exp i(2\phi'_P + \pi - \phi'_N)$$

$$= |F'_N| \exp(i\phi'_N) - |F'_N| \exp(-i\phi'_N) \exp(2i\phi'_P)$$

$$= F'_N - \tilde{F}'_N \exp(2i\phi'_P). \tag{24}$$

The first term leads to the correct structure F'_N, while the second term leads to the convolution of the *negative* inverse structure with the phase-squared structure of the P atoms. The peaks due to the latter can be expected to be distributed and will therefore form only a background. A comparison of the above with the double-phased Fourier in the isomorphous-replacement case, (9) of Ch. 9, brings out one important feature. While the term $\tilde{F}'_N \exp(2i\phi'_P)$ is positive in the isomorphous-replacement case, the corresponding term $-\tilde{F}'_N \exp(2i\phi'_P)$ appears with a negative sign in (24). The background peaks in the present case are therefore negative. The identification of the true peaks would therefore be expected to be easier in the present case of the double-phased anomalous-dispersion synthesis.

The complementary nature of the isomorphous-replacement and double-phased anomalous-dispersion syntheses is a particularly noteworthy feature. We shall see that this feature persists in various other types of syntheses which may be obtained from isomorphous crystal and anomalous-dispersion data, which can be utilized for various purposes (see Ch. 13).

Centrosymmetric anomalous-dispersion group

The situation when the anomalously scattering atoms form a centrosymmetric constellation, while the structure as a whole is noncentrosymmetric, is of particular interest. In this case, $\phi_P' = 0$ or π, so that $2\phi_P'$ is zero in either case. Thus (24) becomes equal to

$$F_N' - \tilde{F}_N' = 2i\,|F_N'|\sin\phi_N'. \tag{25}$$

It is obvious from this that we will obtain in the double-phased anomalous-dispersion synthesis the structure F_N' and its inverse $\tilde{F}_N'$, the latter having *negative strength*. The point of inversion is obviously the center of symmetry of the P group. The similarity between the above case and the corresponding case with isomorphous crystals is at once obvious. Thus in both duplication of peaks occurs about the center of inversion of the P group of atoms. However, there is a distinction, in that, while the inverse peaks are positive in the isomorphous-replacement case, they are negative in the anomalous-dispersion case.[37]

Thus, in a sense, the anomalous-dispersion case is superior to the isomorphous-replacement case, since the correct structure is the one corresponding to the positive peaks, which can be readily distinguished from the inverse structure which is negative. Further, if the necessary precautions are taken in indexing the reflections,[27,39] the structure so obtained also yields the absolute configuration.

Condition for the occurrence of Bijvoet differences

Equation (9) shows that $\Delta I(\mathbf{H}) = 4\,|F'(\mathbf{H})|\,|F''(\mathbf{H})|\cos\psi$, in which $F_P''(\mathbf{H})$ of (9) is written as $F''(\mathbf{H})$ to take account of the fact that all atoms may have anomalous-dispersion effects. This can be expressed in terms of the components of the Argand vectors $F'(\mathbf{H})$ and $F''(\mathbf{H})$. Thus, we may write

$$F'(\mathbf{H}) = A'(\mathbf{H}) + iB'(\mathbf{H}), \tag{26}$$

where

$$A'(\mathbf{H}) = \sum_{j=1}^{N} f_{Nj}' \cos(2\pi\mathbf{H}\cdot\mathbf{r}_{Nj}) \tag{27a}$$

$$B'(\mathbf{H}) = \sum_{j=1}^{N} f_{Nj}' \sin(2\pi\mathbf{H}\cdot\mathbf{r}_{Nj}). \tag{27b}$$

Similarly, we define

$$F''(\mathbf{H}) = i[A''(\mathbf{H}) + iB''(\mathbf{H})], \tag{28}$$

where

$$A''(\mathbf{H}) = \sum_{j=1}^{N} f''_{Nj} \cos (2\pi\mathbf{H} \cdot \mathbf{r}_{Nj}), \tag{29a}$$

$$B''(\mathbf{H}) = \sum_{j=1}^{N} f''_{Nj} \sin (2\pi\mathbf{H} \cdot \mathbf{r}_{Nj}). \tag{29b}$$

We thus have

$$F(\mathbf{H}) = [A'(\mathbf{H}) - B''(\mathbf{H})] + i[B'(\mathbf{H}) + A''(\mathbf{H})], \tag{30a}$$

$$F(\overline{\mathbf{H}}) = [A'(\mathbf{H}) + B''(\mathbf{H})] - i[B'(\mathbf{H}) - A''(\mathbf{H})], \tag{30b}$$

from which it can be calculated that

$$\Delta I = |F(\mathbf{H})|^2 - |F(\overline{\mathbf{H}})|^2 = 4[A''(\mathbf{H})B'(\mathbf{H}) - A'(\mathbf{H})B''(\mathbf{H})]. \tag{31}$$

This expression is readily shown to be equivalent to (9) with $P = N$. It follows from (31) that $\Delta I = 0$ if $A''(\mathbf{H})$ and $B''(\mathbf{H})$ are zero, which is automatically true for all reflections $\mathbf{H}$ if f''_{Nj} is zero for all the atoms in the crystal. This is the familiar condition for Friedel's law to be valid, which is that there should be no absorption components for the atomic scattering factors.

On the other hand, the right-hand side of (31) can also be zero under another condition, namely when

$$\frac{A''(\mathbf{H})}{A'(\mathbf{H})} = \frac{B''(\mathbf{H})}{B'(\mathbf{H})}. \tag{32}$$

If now k_{Nj} ($= f''_{Nj}/f'_{Nj}$) is the same for all the atoms in the crystal, this condition is automatically satisfied for *every* reflection, and therefore, no Bijvoet differences can be observed for a crystal composed of a *single* species of atom, even though the atom may have anomalous dispersion (see below for possible exceptions to this rule).

On the other hand, although (32) may be satisfied for some reflections, it is not true for all if some k_{Nj} are different from the others, because the cosine and sine coefficients in (27) and (29) vary from reflection to reflection. Hence, failure of Friedel's law will be observed in any crystal in which only some, but not all, atoms exhibit anomalous effects. Even if all have some $\Delta f''$, provided k_{Nj} is not the same for all the atoms, Bijvoet differences should be observed.

Condition (32) implies that $\cos \psi = 0$, or that F''_P is perpendicular to F' in Fig. 3, which again leads to the same conclusions as above. This discussion indicates the possibility of observing Bijvoet differences (and hence of determining the absolute configuration) of crystals having a single species of atoms, provided the constancy of k ($= f''/f'$) is not there for the different atoms in a unit cell. This could be possible, for instance, by the occurrence of anisotropic vibration effects, or difference in the isotropic thermal parameters B_j, for the various atoms, as has been pointed out by Chandrasekharan.[3] If there is more than one atom in the unit cell, these effects could make the value of k for the two atoms different for a particular reflection, and hence (32) would not be satisfied, and the right-hand side of (31) would not be zero.

Determination of $\Delta f''$

We have considered in the preceding sections the use of observations on Bijvoet differences (which arise essentially from the occurrence of $\Delta f''$ for some of the atoms in the crystal structure) for determining the phases of the reflections and hence the locations of the atoms in the unit cell. Once the structure is actually determined, the process can be reversed, and (9), or its equivalent form (31), can be used to determine the value of $\Delta f''$. In this connection, still another equivalent form of (31), as worked by Zachariasen,[44] which contains explicitly the anomalous ratios k_j $(=f_j''/f_j')$ of the various atoms, is particularly useful. Substituting from (27) and (29) in (31), we obtain

$$\Delta I = 2 \sum_j \sum_l (k_j - k_l) f_j' f_l' \sin(\phi_l - \phi_j), \tag{33}$$

where

$$\phi_j = 2\pi \mathbf{H} \cdot \mathbf{r}_j. \tag{34}$$

Chlorine and barium

The first attempt to determine $\Delta f''$ in this way was made by Parthasarathy,[21,22] who measured this for chlorine[21,22] and barium.[22] For the former purpose, he used crystals of tyrosine hydrochloride, and for the latter, those of barium glucose orthophosphate. He measured the Bijvoet ratios of a number of reflections for Cu $K\alpha$ radiation using a Geiger counter, from which the Bijvoet differences ΔI were obtained, since the absolute values of the structure amplitudes could be calculated. On calculating the value of $\Delta f''$ of chlorine from these, a value of 0.66 was obtained for small $(\sin \theta)/\lambda$. For larger values of $(\sin \theta)/\lambda$, there were large fluctuations in the experimentally determined value of $\Delta f''$ which could be attributed mainly to the errors in the atomic coordinates. (The structure had been determined only in two projections). However, the mean value of $\Delta f''$ did not show a significant decrease with increase of $\sin \theta$, as is to be expected from theory. The experimental value of 0.67 ± 0.21 agrees with the value of 0.7, as given in the *International tables for x-ray crystallography* (1962). The value of $\Delta f''$ obtained for barium was 9.9 ± 2.5, as compared with 8·9 given in the *International tables*.

Iodine

More recently, Hall and Maslen[11] determined $\Delta f''$ of iodine by photographic methods, using the structure of methyl melaleucate iodoacetate and Cu$K\alpha$ radiation. In this case also the experimentally determined value, 6.3, is in reasonable agreement with the theoretical value 7.2 (*International tables*), which shows that the value of $\Delta f''$ is not affected by the presence of more than one anomalous scatterer in the unit cell, a conclusion also arrived at by Parthasarathy.[21,22] However, in refining their structure Hall and Maslen did not include anomalous-dispersion effects, which are important in this case because they are large.

Silicon and oxygen

Perhaps the most accurate determinations of $\Delta f''$ have been made by Zachariasen[44] for quartz, using counter methods. In all the cases mentioned above, the light atoms were assumed to have negligible values of $\Delta f''$. How-

Table 8 Bijvoet-ratio data for quartz

h	k	l	$\|F_{obs}\|$	Exp. %	Calc. %	h	k	l	$\|F_{obs}\|$	Exp. %	Calc. %
1	1	0	17.84	3	6	3	1	$\bar{2}$	16.49	−2	−3
		1	9.64	21	20			3	6.53	11	8
		2	24.31	3	1			$\bar{3}$	7.48	0	0
		3	11.62	−3	−5			4	9.65	5	3
		4	17.13	−2	−4			$\bar{4}$	5.68	5	5
		5	11.75	−3	−5			5	5.68	1	−3
		6	6.24	0	−2			$\bar{5}$	8.13	2	6
2	1	0	4.29	22	22	3	2	0	10.26	2	0
		1	16.63	5	4			1	1.87	−1	−3
		$\bar{1}$	18.00	0	−1			$\bar{1}$	15.47	0	−2
		2	10.94	8	7			2	9.25	5	1
		$\bar{2}$	18.96	2	−1			$\bar{2}$	2.80	10	11
		3	15.94	2	1			3	0.74	51	56
		$\bar{3}$	8.32	−8	−8			$\bar{3}$	1.16	−41	−44
		4	14.39	1	−2			4	5.95	−1	−2
		$\bar{4}$	3.32	−33	−25			$\bar{4}$	15.44	0	2
		5	4.11	−18	−19	4	1	0	3.75	5	3
		$\bar{5}$	12.46	−2	−1			1	13.59	4	4
		6	3.36	−15	−19			$\bar{1}$	5.35	−2	−1
		$\bar{6}$	9.07	0	3			2	6.46	0	−1
2	2	0	18.70	−2	−4			$\bar{2}$	8.08	−3	−4
		1	9.89	0	−2			3	4.53	−3	−3
		2	1.96	−37	−35			$\bar{3}$	13.63	0	0
		3	13.84	6	3	3	3	0	11.51	3	0
		4	8.70	5	3			1	8.76	6	5
		5	9.06	3	3			2	1.92	−30	−30
3	1	0	17.86	−1	−2	4	2	0	11.88	1	3
		1	7.60	7	4			1	3.99	0	2
		$\bar{1}$	12.74	−5	−7			$\bar{1}$	5.96	5	9
		2	10.61	10	8						

ever, in the case of SiO_2, Zachariasen considered the absorption components of the scattering factors of both Si and O atoms, and determined the best values of both of these. He obtained $\Delta f''_{Si} = 0.31$ and $\Delta f''_O = 0.028$. The accuracy of the determination may be judged from the comparison between the observed and calculated values of the Bijvoet ratio in Table 8. The mean value of the Bijvoet ratio is only 7 per cent, but the individual values go up to 50 per cent. The value of the mean, calculated from theory of Parthasarathy[23] neglecting the oxygen contribution, is also 7 per cent.

The experimental values of $\Delta f''_{Si}$ and $\Delta f''_O$ obtained in the above study are rather smaller than those given in the *International tables* ($\Delta f''_{Si} = 0.4$, $\Delta f''_O = 0.1$), but are only slightly smaller than those obtained from (35), connecting $\Delta f''$ with the atomic absorption coefficient μ_a:

$$\mu_a = \frac{2e^2\lambda}{mc^2}\,\Delta f'', \tag{35}$$

namely[44] $\Delta f''_{Si} = 0.33$, $\Delta f''_O = 0.035$.

Aluminum and gallium

Following the work of Zachariasen, Marezio[16,17] determined the values of $\Delta f''_{Al}$ and $\Delta f''_{Ga}$ for $CuK\alpha$ radiation using the two crystals γ $LiAlO_2$ and $LiGaO_2$. In the former case, both $\Delta f''_{Al}$ and $\Delta f''_O$ were determined, which were found to be 0.22 ± 0.02 and 0.028 ± 0.005, respectively. As with quartz, these are significantly lower than the values 0.3 and 0.1 given in the *International tables*, but agree better with the values 0.25 and 0.035 calculated from the atomic absorption coefficients.

In the case of $LiGaO_2$, the values of $\Delta f''_{Ga}$ ($= 0.90$) and $\Delta f''_O$ ($= 0.035$) were calculated from the atomic absorption coefficients, and the calculated Bijvoet ratios obtained using these data agree well with observation. Assuming the value of 0.028 for $\Delta f''_O$ as determined for $LiAlO_2$, the best value of $\Delta f''_{Ga}$ was deduced to be 0.89 ± 0.03. In this case, this agrees well with the value 0.9 given in the *International tables*.

Refinement of structures in the presence of anomalous-dispersion effects

The problem of treating anomalous-dispersion effects in crystal-structure refinement has been assuming importance in view of the increased accuracy that is possible nowadays in the experimental measurement of intensities and the development of refined techniques for treating the errors arising from extinction and similar effects. In most cases where the accuracy is not very high (as in many photographic investigations) it would be sufficient to take

$I(\mathbf{H})$ as $[I(\mathbf{H}) + I(\overline{\mathbf{H}})]/2$, and take this to be equal to $|F(\mathbf{H})|^2$, neglecting $|F''(\mathbf{H})|^2$. Since, strictly, $|F'(\mathbf{H})|^2 = [I(\mathbf{H}) + I(\overline{\mathbf{H}})]/2 - |F''_P(\mathbf{H})|^2$, it is indeed possible to calculate the correct value from the known positions of the anomalous scatterers.

The need for taking into account the correction for the absorption part was pointed out by Templeton,[42] later amplified by Patterson.[26] In particular, failure to take into account the F'' correction and performing the refinement using only partial data (such as using only the measured values of $I(\mathbf{H})$ over a half of the reciprocal sphere) can lead to considerable inaccuracy in the coordinates of the light atoms in certain polar space groups.[6,26,43] For example in the case of the structure of thorium nitrate pentahydrate,[43] shifts of the order of 0.05 Å were observed in the coordinates. Even the above procedure (i.e., correcting for F'' alone) gives only $|F'(\mathbf{H})|^2$ which yields a structure corresponding to the dispersion component f'_j of the atoms in the unit cell, at the particular wavelength used, but not corresponding to f^0_j, the normal component independent of wavelength. Patterson[26] suggested a method of refinement of the structure from the observed data which would correspond to the electron-density distribution $\rho^0(\mathbf{r})$ corresponding to the normal scattering factors. Thus, assuming that there is only one type of anomalous scatterers in the unit cell, (30a) and (30b) can be written as

$$F_N(\mathbf{H}) = (A^0_N + gA^0_P - kB^0_P) + i(B^0_N + gB^0_P + kA^0_P) \tag{36a}$$

$$F_N(\overline{\mathbf{H}}) = (A^0_N + gA^0_P + kB^0_P) - i(B^0_N + gB^0_P - kA^0_P), \tag{36b}$$

where A^0_N and B^0_N are the real and imaginary components of the total normal vector F^0_N and g and k are defined by

$$g = \frac{\delta f'_P}{f^0_P}, \qquad k = \frac{\Delta f''}{f^0_P}. \tag{37}$$

Proceeding as before, it can be deduced that

$$\left[\frac{I(\mathbf{H}) + I(\overline{\mathbf{H}})}{2}\right] = |F^0_N|^2 + (g^2 + k^2)|F^0_P|^2 + 2g(A^0_N A^0_P + B^0_N B^0_P), \tag{38a}$$

and

$$\left[\frac{I(\mathbf{H}) - I(\overline{\mathbf{H}})}{2}\right] = -2k(A^0_N B^0_P - B^0_N A^0_P). \tag{38b}$$

It is neither $|F'_N|$, $[I(\mathbf{H}) + I(\overline{\mathbf{H}})]/2$, nor $|F(\mathbf{H})|$, but actually the quantity $|F^0_N|$ that should be used in the Fourier synthesis. From (38a), its value could be deduced to be

$$|F^0_N|^2 = I_{\text{mean}} - (g^2 + k^2)|F^0_P|^2 - 2g(A^0_N A^0_P + B^0_N B^0_P). \tag{39}$$

Once having solved the structure fairly accurately, further refinement should involve the quantity $|F_N^0|$ defined by (39), which could be calculated at any particular stage of refinement. In the centrosymmetric case, this reduces to

$$|F_N^0|^2 = |F(\mathbf{H})|^2 - (g^2 + k^2)A_P^{0^2} - 2gA_N^0 A_P^0. \tag{40}$$

The above approach is equivalent, at any particular stage of refinement, to deriving new sets of $|F_N^0|$ obtained as corrections from observed intensity data and may not be advantageous from a practical point of view. Ibers and Hamilton[13] suggested that the same could be achieved by incorporating these corrections in the calculated structure factors in the least-squares program. Thus the Fourier synthesis corresponding to F_N^0 can be defined as a limiting function defined by

$$\rho^0(\mathbf{r}) = \frac{1}{V} \sum_{\mathbf{H}} |F(\mathbf{H})|_{\mathrm{obs}} \left[\frac{A_N^0(\mathbf{H})}{|F(\mathbf{H})|_{\mathrm{calc}}} \cos(2\pi\mathbf{H}\cdot\mathbf{r}) + \frac{B_N^0(\mathbf{H})}{|F(\mathbf{H})|_{\mathrm{calc}}} \sin(2\pi\mathbf{H}\cdot\mathbf{r}) \right]$$
$$+ \frac{1}{V} \sum_{\mathbf{H}} |F(\overline{\mathbf{H}})|_{\mathrm{obs}} \left[\frac{A_N^0(\overline{\mathbf{H}})}{|F(\overline{\mathbf{H}})|_{\mathrm{calc}}} \cos(2\pi\overline{\mathbf{H}}\cdot\mathbf{r}) + \frac{B_N^0(\overline{\mathbf{H}})}{|F(\overline{\mathbf{H}})|_{\mathrm{calc}}} \sin(2\pi\overline{\mathbf{H}}\cdot\mathbf{r}) \right]. \tag{41}$$

It could be readily seen that, in the limit, the above synthesis converges to $\rho^0(r)$. This has the advantage that the observed values of $|F(\mathbf{H})|$ and $|F(\overline{\mathbf{H}})|$ are used without any change throughout the refinement.

In the centosymmetric case, the refinement is equivalent to iteratively performing the synthesis[13]

$$\rho^0(r) = \frac{1}{2V} \sum_{\mathbf{H}} |F(\mathbf{H})|_{\mathrm{obs}} \frac{A_N^0 \cos(2\pi\mathbf{H}\cdot\mathbf{r})}{[(A_N^0 + gA_P^0)^2 + k^2 B_P^{0^2}]^{\frac{1}{2}}}. \tag{42}$$

Physically the above is equivalent to using the component $|F_N^0| = |F(\mathbf{H})|_{\mathrm{obs}}$ $\cos \varepsilon$, where ε is the finite phase shift introduced for any centric reflection by the presence of the anomalous scatterers. As was pointed out by Templeton,[42] this shift could be really large, especially for weak reflections, arising from the opposite signs of the contributions from the P and Q atoms.

The above considerations clearly show that, for accurate analysis, it is desirable to have intensity data collected over the complete sphere of reflection for a noncentrosymmetric crystal and not over a part.[12] This should be possible nowadays with multicircle diffractometers.

References

[1] J. M. Bijvoet. *Phase determination in direct Fourier synthesis of crystal structures.* Proc. Koninkl. Ned. Akad. Wetenschap (B) **52** (1949) 313–314.

[2] J. M. Bijvoet. *Structure of optically active compounds in the solid state.* Nature **173** (1954) 888–891.

[3] K. S. Chandrasekharan. *Possibility of " Bijvoet differences " in noncentrosymmetric structures of elements.* Acta Crystallogr. **A24** (1968) 248–249.

[4] D. Coster, K. S. Knol, and J. A. Prins. *Unterschiede in der intensitat der Rontgenstrahlenreflexion an den berden III — Flachen der Zinkblende.* Z. Phys. **63** (1930) 345–369.

[5] Don T. Cromer. *Anomalous dispersion corrections computed from self-consistent field relativistic Dirac-Slater wave functions.* Acta Crystallogr. **18** (1965) 17–23

[6] D. W. J. Cruickshank and W. S. McDonald. *Parameter errors in Polar space groups caused by neglect of anomalous scattering.* Acta Crystallogr. **23** (1967) 9–11.

[7] D. Dale, D. C. Hodgkin, and K. Venkatesan. *The determination of crystal structure of Factor V1a.* In "Crystallography and crystal perfection" Ed. G. N. Ramachandran, (Academic Press, London, 1963). 237–242.

[8] Carol H. Dauben and David H. Templeton. *A table of dispersion corrections for x-ray scattering of atoms.* Acta Crystallogr. **8** (1955) 841–842.

[9] T. J. H. Geurtz. *Application of anomalous scattering method in the structure determination of cytisine hydrobromide.* Ph.D. Thesis, Rijksuniversiteit, Utrecht (1963).

[10] S. R. Hall and E. N. Maslen. *The determination of the crystal structure of methyl melaleucate iodoacetate.* Acta Crystallogr. **18** (1965) 265–279.

[11] S. R. Hall and E. N. Maslen. *An experimental determination of $\Delta f''$ for iodine.* Acta Crystallogr. **20** (1966) 383–389.

[12] J. A. Ibers. *Some advantages of a complete data set.* Acta Crystallogr. **22** (1967) 604–605.

[13] J. A. Ibers and W. C. Hamilton. *Dispersion corrections and crystal structure refinements.* Acta Crystallogr. **17** (1963) 781–782.

[14] R. A. Jacobson, J. A. Wunderlich, and W. N. Lipscomb. *The crystal and molecular structure of cellobiose.* Acta Crystallogr. **14** (1961) 598–607.

[15] R. W. James. *The crystalline state. Vol. II. The optical principles of the diffraction of x-rays.* Cornell University Press (1965).

[16] M. Marezio. *The crystal structure and anomalous dispersion of γ-LiAlO$_2$.* Acta Crystallogr. **19** (1965) 396–400.

[17] M. Marezio. *Anomalous dispersion in LiGaO$_2$.* Acta Crystallogr. **19** (1965) 284–285.

[18] Y. L. Oh and E. N. Maslen. *The structure of davallol iodoacetate.* Acta Crystallogr. **20** (1966) 852–864.

[19] Y. Okaya and R. Pepinsky. *New formulation and solution of the phase problem in x-ray analysis for noncentrosymmetric crystals containing anomalous scatterers.* Phys. Rev. **103** (1956) 1645–1657.

[20] Y. Okaya, Y. Saito, and R. Pepinsky. *New method in x-ray crystal structure determination involving the use of anomalous dispersion.* Phys. Rev. **98** (1955) 1858–1859.

[21] R. Parthasarathy. *Determination of the anomalous scattering factor $\Delta f''$ for chlorine.* Acta Crystallogr. **15** (1962) 41–46.

[22] R. Parthasarathy. *Studies on the anomalous dispersion of x-rays.* Ph.D. Thesis (University of Madras) 1962.

[23] S. Parthasarathy. *Expectation value of the Bijvoet ratio*. Acta Crystallogr. **22** (1967) 98–103.

[24] S. Parthasarathy. *Probability distribution of the phases in a crystal with heavy atoms. II. Noncentrosymmetric crystal. Probability distribution of the phase angle.* Acta Crystallogr. **18** (1965) 1028–1033.

[25] S. Parthasarathy, G. N. Ramachandran, and R. Srinivasan. *A weighting function to resolve the two-fold ambiguity in the phase determined by the anomalous dispersion method in crystal structure analysis.* Curr. Sci. (India) **33** (1964) 637–638.

[26] A. L. Patterson. *Treatment of anomalous dispersion in x-ray diffraction data.* Acta Crystallogr. **16** (1963) 1255–1256.

[27] A. F. Peerdeman and J. M. Bijvoet. *The indexing of reflections in investigations involving the use of anomalous scattering effect.* Acta Crystallogr. **9** (1956) 1012–1015.

[28] A. F. Peerdeman, A. J. Van Bommel, and J. M. Bijvoet. *Determination of the absolute configuration of optically active compounds by means of x-rays.* Proc. Koninkl. Ned. Akad. Wetenschap (B)**54** (1951) 16–19.

[29] R. Pepinsky and Y. Okaya. *Comparison of two procedures for solution of noncentric crystal structures utilizing anomalous dispersion.* Phys. Rev. **108** (1957) 1231–1232.

[30] S. W. Peterson. *Anomalous x-ray scattering at wavelengths far from an absorption edge.* Nature **176** (1955) 395.

[31] D. C. Phillips. *The crystal and molecular structure of ephedrine hydrochloride.* Acta Crystallogr. **7** (1954) 159–165.

[32] G. N. Ramachandran. *Fourier syntheses for partially known crystal structures.* In "Advanced methods of crystallography". Ed. G. N. Ramachandran (Academic Press, London, 1963). 25–65.

[33] G. N. Ramachandran and R. R. Ayyar. *Fourier syntheses for feeding in isomorphous and anomalous dispersion data.* In "Crystallography and crystal perfection" Ed. G. N. Ramachandran (Academic Press, London, 1963) 25–41.

[34] G. N. Ramachandran and S. Parthasarathy. *Anomalous dispersion method. Its power for protein structure analysis.* Science **150** (1965) 212–214.

[35] G. N. Ramachandran and S. Raman. *A new method for the structure analysis of noncentrosymmetric crystals.* Curr. Sci. (India) **25** (1956) 348–351.

[36] G. N. Ramachandran and R. Srinivasan. *A new statistical test for distinguishing between centrosymmetric and noncentrosymmetric structures.* Acta Crystallogr. **12** (1959) 410–411.

[37] S. Raman. *Anomalous dispersion method of determining structure and absolute configuration of crystals.* Proc. Ind. Acad. Sci. **A47** (1958) 1–11.

[38] S. Raman. *Determination of the structure and absolute configuration of L(+) lysine hydrochloride dihydrate by the anomalous dispersion method.* Z. Kristallogr. **111** (1959) 301–317.

[39] S. Ramaseshan. *The use of anomalous scattering in crystal structure analysis.* In "Advanced methods of crystallography". Ed. G. N. Ramachandran (Academic Press, London, 1963) 67–95.

[40] G. D. Rieck and Caroline H. MacGillavry. *International tables for x-ray crystallography*, Vol. 3 (Kynoch Press, Birmingham, England, 1960).

[41] G. A. Sim. *A note on the determinations of phases by anomalous dispersion*. Acta Crystallogr. **17** (1964) 1072–1073.

[42] D. H. Templeton. *X-ray dispersion effects in crystal structure determinations*. Acta Crystallogr. **8** (1955) 842–843.

[43] T. Ueki, A. Zalkin, and D. H. Templeton. *Crystal structure of thorium nitrate pentahydrate by x-ray diffraction*. Acta Crystallogr. **20** (1966) 836–841.

[44] W. H. Zachariasen. *Dispersion in quartz*. Acta Crystallogr. **18** (1965) 714–716.

12

Anomalous dispersion effect: Fourier syntheses

Syntheses of the alpha and beta classes with anomalous-dispersion data

The alpha and beta classes of syntheses for the case when the anomalous-dispersion effect is present are readily formulated. We shall consider the general case of a noncentrosymmetric crystal containing N atoms, P of which are anomalous scatterers. We shall also assume that the P atoms are not necessarily alike.

Alpha anomalous-dispersion synthesis

The alpha anomalous-dispersion (α_{an}) synthesis is defined[9,11] as

$$\alpha_{an} = [\tfrac{1}{2}\,\Delta\,|F_N|^2 - (F'_P\tilde{F}''_P + \tilde{F}'_P F''_P)]F''_P \tag{1}$$

The reason for taking this particular form will be clear later—see (5).

All the quantities on the right-hand side of (1) are available, since they involve quantities which are either measured $(\Delta\,|F_N|^2)$, or calculable, since the P atoms are assumed to be known. We may use the relations

$$F_N(\mathbf{H}) = F_Q(\mathbf{H}) + F'_P(\mathbf{H}) + F''_P(\mathbf{H}) \tag{2a}$$

$$F_N(\overline{\mathbf{H}}) = F_Q(\overline{\mathbf{H}}) + F'_P(\overline{\mathbf{H}}) + F''_P(\overline{\mathbf{H}}). \tag{2b}$$

Further we have

$$F_Q(\overline{\mathbf{H}}) = \tilde{F}_Q(\mathbf{H}), \qquad F'_P(\overline{\mathbf{H}}) = \tilde{F}'_P(\mathbf{H}), \qquad F''_P(\overline{\mathbf{H}}) = -\tilde{F}''_P(\mathbf{H}). \tag{3}$$

201

Using these, we obtain, after some simplification,

$$\Delta |F_N|^2 = |F_N(\mathbf{H})|^2 - |F_N(\overline{\mathbf{H}})|^2,$$
$$= 2(F_P' \tilde{F}_P'' + \tilde{F}_P' F_P'') + 2(F_Q \tilde{F}_P'' + \tilde{F}_Q F_P'') \tag{4}$$

where $F_P' = F_P'(\mathbf{H})$, $\tilde{F}_P'' = \tilde{F}_P''(\mathbf{H})$, etc., and the symbol $\mathbf{H}$ within brackets is omitted hereafter for convenience. Thus, α_{an} reduces to

$$\alpha_{an} = F_Q |F_P''|^2 + \tilde{F}_Q F_P''^2. \tag{5}$$

A comparison of (5) with the α_{is} synthesis (10) of Ch. 10 shows that the two are closely similar. Instead of $(\delta F_P)^2$ and $|\delta F_P|^2$ appearing in α_{is}, $F_P''^2$ and $|F_P''|^2$ occur in α_{an}. The synthesis $F_P''^2$ has the nature of the negative squared structure of the P group and $|F_P''|^2$ is similar to the Patterson of the P group. In fact, if the P atoms are alike and the coefficients are all multiplied by a factor $(f_P/\Delta f_P'')$ we will obtain exactly the negative squared structure and the Patterson of the P group, respectively, for these two syntheses.

With these modifications, our earlier discussions of α_{is} are applicable to the α_{an} synthesis also. It contains the structure F_Q arising out of the first term in (5), while the second term is equivalent to a convolution of the inverse structure $\tilde{F}_Q$ with the squared structure of F_P''. More generally,

$$F_P''^2 = [i \sum f_{Pj}'' \exp (2\pi i \mathbf{H} \cdot \mathbf{r}_j)]^2$$
$$= -\sum \sum f_{Pj}'' f_{Pk}'' \exp [2\pi i \mathbf{H} \cdot (\mathbf{r}_{Pj} + \mathbf{r}_{Pk})], \tag{6}$$

so that the peaks in this synthesis occur at locations $\mathbf{r}_{Pj} + \mathbf{r}_{Pk}$ of the squared structure of the P atoms, but have negative strengths, of magnitude $-f_{Pj}'' f_{Pk}''$. On the other hand,

$$|F_P''|^2 = |i \sum f_{Pj}'' \exp (2\pi i \mathbf{H} \cdot \mathbf{r}_j)|^2$$
$$= +\sum \sum f_{Pj}'' f_{Pk}'' \exp [2\pi i \mathbf{H} \cdot (\mathbf{r}_{Pj} - \mathbf{r}_{Pk})]. \tag{7}$$

Hence, the synthesis $|F_P''|^2$ has positive peaks of strength $f_{Pj}'' f_{Pk}''$ at the Patterson positions $\mathbf{r}_{Pj} - \mathbf{r}_{Pk}$. In view of these, all the peaks in the α_{an} synthesis can be exactly listed (Table 1a). It will be seen that there is no peak strength at the

Table 1   List of peaks in the alpha anomalous-dispersion synthesis
1a, Noncentrosymmetric P group of atoms

Position	Strength	Nature
$\mathbf{r}_{Qk}$	$(\sum f_{Pj}''^2) f_{Qk}$	Unknown atoms
$\mathbf{r}_{Pi} - \mathbf{r}_{Pj} + \mathbf{r}_{Qk}$ $(i \neq j)$	$f_{Pi}'' f_{Pj}'' f_{Qk}$	Positive background
$2\mathbf{r}_{Pj} - \mathbf{r}_{Qk}$	$-2 f_{Pj}'' f_{Qk}$	Negative background
$(\mathbf{r}_{Pi} + \mathbf{r}_{Pj}) - \mathbf{r}_{Qk}$ $(i \neq j)$	$-f_{Pi}'' f_{Pj}'' f_{Qk}$	Negative background

$$1b, \text{ Centrosymmetric } P \text{ group}^\dagger$$

Position	Strength	Nature
$+\mathbf{r}_{Qk}$	$(\sum f_{Pj}''^2)f_{Qk}$	Required Q atoms, positive peaks
$-\mathbf{r}_{Qk}$	$-(\sum f_{Pj}''^2)f_{Qk}$	Inverse Q structure, negative peaks
$\pm 2\mathbf{r}_{pi} + \mathbf{r}_{Qk}$	$f_{pj}^2 f_{Qk}$	Positive background
$\pm 2\mathbf{r}_{pi} - \mathbf{r}_{Qk}$	$-f_{pj}^2 f_{Qk}$	Negative background
$\pm(\mathbf{r}_{pi} \pm \mathbf{r}_{pj}) + \mathbf{r}_{Qk}$	$f_{pi} f_{pj} f_{Qk}$	Positive background
$\pm(\mathbf{r}_{pi} \pm \mathbf{r}_{pj}) - \mathbf{r}_{Qk}$	$-f_{pi} f_{pj} f_{Qk}$	Negative background

† The symbol $p(=P/2)$ is the number of atoms in the asymmetric unit exactly as in Table 2 of Ch. 5.

positions of the known atoms, but that the unknown Q atoms come out as a result of the convolution of the structure F_Q with the structure $|F_P''|^2$, from the interaction of the origin peak of $|F_P''|^2$ with the peaks f_{Qk} at $\mathbf{r}_{Qk}$, giving peaks of strength $(\sum f_{Pj}''^2)f_{Qk}$. On the other hand, the convolution of the structure $F_P''^2$ with the structure $\tilde{F}_Q$ does not, in general, give rise to any large peaks, but a number of small, dispersed peaks.

An exception to the last stated result is the case when the P group of atoms forms a centrosymmetric constellation. In this case, clearly $|F_P''|^2 = -F_P''^2$, and the α_{an} synthesis becomes

$$\alpha_{an} = |F_P''|^2(F_Q - \tilde{F}_Q). \tag{8}$$

The crystal as a whole, and hence the Q atoms, are obviously noncentro-symmetric, and therefore the peaks that arise in the α_{an} synthesis in the special case, when the P atoms alone are centrosymmetric, are as shown in Table 1b. It will be seen that there is a concentration of peak strength at the atomic positions $\mathbf{r}_{Qk}$ (by the interaction of the origin peak of $|F_P''|^2$ with the structure F_Q) of strength $(\sum f_{Pj}''^2)f_{Qk}$, but in addition a set of negative peaks of equal strength appear at the inverse positions $-\mathbf{r}_{Qk}$, by the convolution of $F_P''^2 (= -|F_P''|^2)$ with F_Q. There are also a number of dispersed peaks, of both positive and negative strengths. Thus, in the case when the P atoms have a center of symmetry, the α_{an} synthesis gives the Q atoms duplicated by its negative inverse. We shall not consider this further here, but discuss it in connection with the corresponding case of the β_{an} synthesis.

Beta anomalous-dispersion synthesis

The beta anomalous-dispersion synthesis (β_{an}) may be defined in a manner similar to the α_{an} synthesis. It is given by

$$\beta_{an} = \frac{\alpha_{an}}{|F_P''|^2} = \frac{[\frac{1}{2}\Delta|F_N|^2 - (F_P'\tilde{F}_P'' + \tilde{F}_P'F_P'')]}{|F_P''|} \exp(i\phi_P''). \tag{9}$$

On simplification, this takes the elegant simple form

$$\beta_{an} = F_Q + \tilde{F}_Q \frac{F_P''}{\tilde{F}_P''} \tag{10}$$

$$= F_Q + \tilde{F}_Q \exp(2i\phi_P''). \tag{11}$$

Here again the close similarity of β_{an} with the β_{is} synthesis $[= F_Q + \tilde{F}_Q \exp (2i\phi_P)$, cf. (13) of Ch. 10] is obvious. The only difference is that, instead of ϕ_P in β_{is}, ϕ_P'' appears in β_{an}. This is an important difference whose consequence has interesting applications. The β_{an} synthesis has the structure F_Q in it, i.e., it reveals the unknown Q atoms in their correct location with their correct strengths, and contains, in addition only the peaks of the inverse structure, $\tilde{F}_Q$, convolved with the phase-squared structure $\exp(2i\phi_P'')$ of the P group, which, in general, leads to a small, negative background. The principal peaks expected in this β_{an} synthesis and their strengths are shown in Table 2. It will

Table 2   List of principal peaks in the beta anomalous synthesis
2a, Noncentrosymmetric P group

Position	Strength	Nature
$\mathbf{r}_{Qk}$	f_{Qk}	Require Q atoms
$2\mathbf{r}_{Pj} - \mathbf{r}_{Qk}$	$-2f_{Pj}''f_{Qk}$	Negative background
$\mathbf{r}_{Pi} + \mathbf{r}_{Pj} - \mathbf{r}_{Qk}$	$-[f_{Pi}''f_{Pj}''/(\sum f_{Pj}''^2)]f_{Qk}$	Negative background

2b, Centrosymmetric P group

Position	Strength	Nature
$\mathbf{r}_{Qk}$	f_{Qk}	Required Q atoms
$-\mathbf{r}_{Qk}$	$-f_{Qk}$	Negative inverse of the Q structure
	(No other background at all)	

be noticed that the β_{an} synthesis is exactly equivalent to the double-phased anomalous Fourier[10] [(24) of Ch. 11] except that it involves the Q atoms and not the N atoms. The P atoms are completely absent in the β_{an} synthesis, as in the α_{an} synthesis.

In the special case when all the anomalous scatterers are alike, the Argand vector F_P'' is perpendicular to F_P', and the term $(F_P'\tilde{F}_P'' + \tilde{F}_P'F_P'')$ in (9) vanishes. Also the phase angle $\phi_P'' = \phi_P' + \pi/2$. The coefficient of the β_{an} synthesis reduces to

$$\beta_{an} = \frac{1}{2}\frac{\Delta|F_N|^2}{|F_P''|} \exp\left[i\left(\phi_P + \frac{\pi}{2}\right)\right]$$

$$= \frac{i}{2} \frac{\Delta |F_N|^2}{|F_P''|} \exp (i\phi_P). \tag{12}$$

Correspondingly, (11) becomes

$$\beta_{an} \doteq F_Q - \tilde{F}_Q \exp (2i\phi_P) \tag{13}$$

The fact that the phase of F_P'' is $\pi/2$ ahead of that of F_P' has thus made the second term contain a factor $-\exp (2i\phi_P)$ instead of $\exp (2i\phi_P'')$. On comparing (13) with the β_{is} synthesis [(13) of Ch. 10], it is seen that the background peaks arising from $-\tilde{F}_Q \exp (2i\phi_P)$ in the β_{an} synthesis are equal and opposite to the background peaks in the β_{is} synthesis. A combination of the two will make the background peaks cancel each other (this is discussed more in detail in Ch. 13).

Centrosymmetric P Group. If the P group of atoms is centrosymmetric, we have the result that $\exp (2i\phi_P) \equiv 1$ for all reflections **H**, so that (11) becomes

$$\beta_{an} = F_Q - \tilde{F}_Q \tag{14}$$

In this case, the β_{an} synthesis reveals the structure of the Q atoms, associated with its negative inverse structure, exactly as in the α_{an} synthesis, without, however, any other background. We have seen in Ch. 10 that the β_{is} synthesis gives the structure F_Q associated with the structure $+\tilde{F}_Q$. Once again it is seen that the unwanted peaks in β_{is} and β_{an} occur at exactly the same positions, but with equal strengths of opposite signs. So, in the special case also, a combination of β_{is} and β_{an} suppresses the spurious peaks of the inverse structure, revealing only the correct structure of the Q atoms.

In both the general case and in the special case of a centrosymmetric P group, the positive peaks in the β_{an} synthesis reveal the structure in its absolute configuration (provided the data have been properly indexed, as mentioned in Ch. 11). If the P group is noncentrosymmetric, its absolute configuration itself has to be properly used. It will be found that, if the wrong absolute configuration of the P group is used, then the β_{an} synthesis will not be interpretable, while, if that of the P group is taken correctly, the synthesis reveals the unknown Q atoms in their correct absolute configuration.

Test of the beta anomalous-dispersion synthesis

The method of using both of the ambiguous phases as determined from anomalous-dispersion data, which has been shown to be equivalent to the β_{an} synthesis, was tested by Kartha.[2] The structure was revealed with comparatively little background.

The method of directly using the data to calculate the β_{an} synthesis was tested by Ramachandran and Ayyar.[8] This was carried out for the same model as was used for testing the β_{is} synthesis described in Ch. 10. The only alterations introduced were that the replaceable atoms in one of the crystals were assumed to be anomalous scatterers. They were taken to be all of the same type, so that (13) applies to this case. The synthesis is shown in Fig. 1.

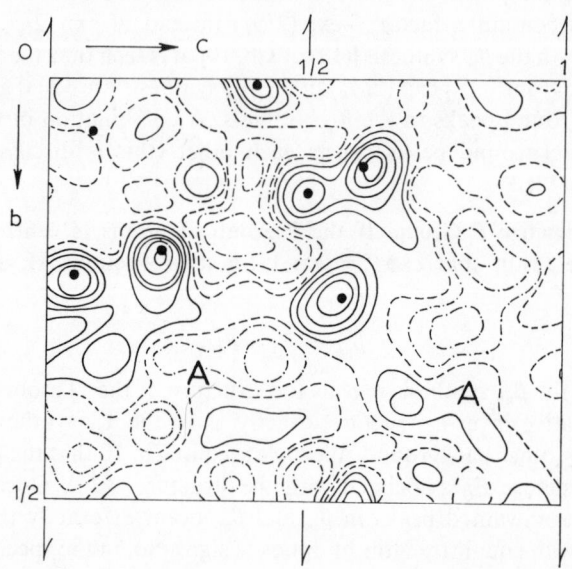

Fig. 1. The β_{an} synthesis for a hypothetical structure. The positions marked A represent the known P atoms (anomalous scatterers) and the dots the unknown Q atoms. Contours are at intervals of 4 $e\text{Å}^{-2}$. Positive contours are continuous lines, negative have dots and dashes and zero contour is dashed.

The required unknown Q atoms are shown by dots and the anomalous scatterers (P), whose positions were used, are shown by the symbol A. It may be seen that all the six unknown atoms are revealed quite conspicuously against a background which is mostly negative. The known atoms are also suppressed as expected from theory.

The Patterson function in the presence of anomalous dispersion[3-7]

So far, we have considered in this chapter and the earlier one how the anomalous-dispersion effect can be used for the determination of phases, and also how a Fourier type of synthesis can be developed for this purpose. We shall now consider another aspect of the same problem, namely the determination of the structure via the Patterson function. This requires a study of the nature of the Patterson function in the presence of anomalous dispersion, and the problem of how the structure could be extracted from it. This method of approach has been largely developed by Pepinsky and co-workers.[3-7]

As discussed in Ch. 2, the normal Patterson function can be written as

$$\mathscr{P}(\mathbf{r}) = \frac{1}{V} \sum_{\mathbf{H}} |F(\mathbf{H})|^2 \exp(-2\pi i \mathbf{H} \cdot \mathbf{r}) \tag{15a}$$

Since $|F(\mathbf{H})|^2 = |F(\overline{\mathbf{H}})|^2$, this may be written in the form

$$\mathscr{P}(\mathbf{r}) = \frac{1}{V} \sum_{\mathbf{H}} |F(\mathbf{H})|^2 \cos(2\pi \mathbf{H} \cdot \mathbf{r}),$$

or

$$\mathscr{P}(\mathbf{r}) = \frac{2}{V} \sum_{\mathbf{H}}' |F(\mathbf{H})|^2 \cos(2\pi \mathbf{H} \cdot \mathbf{r}), \tag{15b}$$

where the prime over the summation denotes that the summation is over one-half the reciprocal sphere (with half weight for the term with $\mathbf{H} = 0$). In the presence of anomalous dispersion, because of the nonequality of $|F(\mathbf{H})|^2$ and $|F(\overline{\mathbf{H}})|^2$, the form (15b) is not valid, and (15a) has to be used. This takes the form

$$\mathscr{P}(\mathbf{r}) = \frac{1}{V} \sum_{\mathbf{H}}' |F(\mathbf{H})|^2 \exp(-2\pi i \mathbf{H} \cdot \mathbf{r})$$
$$+ \frac{1}{V} \sum_{\mathbf{H}}' |F(\overline{\mathbf{H}})|^2 \exp(2\pi i \mathbf{H} \cdot \mathbf{r}). \tag{16}$$

We shall consider the general case when each one of the N atoms may have anomalous-dispersion components $\delta f'$ and $\Delta f''$. The case when some of them are normal scatterers is readily deduced by taking these as zero for those particular atoms. We have

$$|F_N(\mathbf{H})|^2 = |F_N'|^2 + |F_N''|^2 + (F_N' \tilde{F}_N'' + \tilde{F}_N' F_N''), \tag{17a}$$

$$|F_N(\overline{\mathbf{H}})|^2 = |F_N'|^2 + |F_N''|^2 - (F_N' \tilde{F}_N'' + \tilde{F}_N' F_N''), \tag{17b}$$

where

$$= F_N'(\mathbf{H}) \quad \text{and} \quad F_N'' = F_N''(\mathbf{H}). \tag{18}$$

Using (17a) and (17b), we have

$$\mathscr{P}(\mathbf{r}) = \frac{2}{V} \sum{}' \, (|F_N'|^2 + |F_N''|^2) \cos (2\pi \mathbf{H} \cdot \mathbf{r})$$

$$- \frac{2i}{V} \sum{}' \, (F_N' \tilde{F}_N'' + \tilde{F}_N' F_N'') \sin (2\pi \mathbf{H} \cdot \mathbf{r}). \tag{19}$$

It is readily deduced, using (17), that the terms within parentheses in (19) appearing before the cosine and sine function are, respectively, half the sum and half the difference of the intensities $I(\mathbf{H})$ and $I(\overline{\mathbf{H}})$. Thus if we use the notation

$$S|F|^2 = |F(\mathbf{H})|^2 + |F(\overline{\mathbf{H}})|^2, \tag{20a}$$

and

$$\Delta|F|^2 = |F(\mathbf{H})|^2 - |F(\overline{\mathbf{H}})|^2, \tag{20b}$$

(19) can be written as

$$\mathscr{P}(\mathbf{r}) = \mathscr{P}_c(\mathbf{r}) - i\mathscr{P}_s(\mathbf{r}), \tag{21}$$

where

$$\mathscr{P}_c(\mathbf{r}) = \frac{1}{V} \sum{}' \, S|F|^2 \cos (2\pi \mathbf{H} \cdot \mathbf{r}) \tag{22a}$$

$$\mathscr{P}_s(\mathbf{r}) = \frac{1}{V} \sum{}' \, \Delta|F|^2 \sin (2\pi \mathbf{H} \cdot \mathbf{r}). \tag{22b}$$

Thus the consequence of the anomalous-dispersion effect is that the Patterson function is no longer real, but has to be formally represented as a complex function, whose real and imaginary components are $\mathscr{P}_c(\mathbf{r})$ and $\mathscr{P}_s(\mathbf{r})$.

The function $\mathscr{P}_c(\mathbf{r})$ is an even function and is the equivalent of the conventional Patterson synthesis. Using (19), it is readily seen that it represents the Patterson corresponding to the sum of $|F_N'|^2$ and $|F_N''|^2$, i.e.,

$$\mathscr{P}_c(\mathbf{r}) = \text{Structure} \, (|F_N'|^2 + |F_N''|^2). \tag{23}$$

Both the terms in (23) lead to peaks at the same positions, namely

$$\mathbf{r}_{Ni} - \mathbf{r}_{Nj}$$

and the strengths are therefore additive, given by

$$(f_{Ni}' f_{Nj}' + f_{Ni}'' f_{Nj}'').$$

Obviously when anomalous dispersion is absent, this reduces to the conventional Patterson

The function $\mathscr{P}_s(\mathbf{r})$, on the other hand, is an odd function of $\mathbf{r}$, and, using (19), it is seen to be equivalent to

$$\mathscr{P}_s(\mathbf{r}) = \text{Structure}\ (F'_N \tilde{F}''_N + \tilde{F}'_N F''_N) \tag{24}$$

It therefore contains peaks at $\mathbf{r}_{Ni} - \mathbf{r}_{Nj}$ and their strengths are $(f'_{Ni} f''_{Nj} - f''_{Ni} f'_{Nj})$ and, since it is a sine synthesis, there is a set of duplicate peaks at $(\mathbf{r}_{Nj} - \mathbf{r}_{Ni})$ with equal strength, but of opposite sign, i.e., of strength $(f'_{Nj} f''_{Ni} - f''_{Nj} f'_{Ni})$. From its very nature, it is obvious that $\mathscr{P}_s(\mathbf{r})$ is nonzero only for a noncentrosymmetric crystal, since otherwise the coefficients $\Delta |F|^2$ are all automatically zero.

The function $\mathscr{P}_s(\mathbf{r})$ is the one that is important from the point of view of extracting the structure using anomalous dispersion data. We shall examine it more closely. For convenience, we assume that, out of a total of N atoms, P atoms are anomalous scatterers, and Q $(=N-P)$ are non-anomalous (normal) scatterers. Also, let us suppose that the P atoms are all alike. The different vectors $\mathbf{r}_{Ni} - \mathbf{r}_{Nj}$ in the function $\mathscr{P}_s(\mathbf{r})$ can then be classified under three categories—namely the $[PP]$, $[PQ]$, and $[QQ]$ type of vectors. For a $[PP]$ type of vector, it is seen that the strength is $(f'_{Pj} f''_{Pk} - f''_{Pj} f'_{Pk})$, which is identically zero. This means that the $[PP]$ vectors are completely absent in the $\mathscr{P}_s(\mathbf{r})$ synthesis. (Obviously this statement is not true if there is more than one type of anomalous scatterer.) Similarly for the $[QQ]$ type of vector, the strength of the peaks is zero, since $f''_Q = 0$. These $[QQ]$ vectors are always completely absent and this statement is true whether the anomalous scatterers are all of one type or not. Consider now the $[PQ]$ type of interaction. The strength of a typical one is $f''_{Pj} f'_{Qk}$ at $\mathbf{r}_{Pj} - \mathbf{r}_{Qk}$ and there is an inverse peak at $\mathbf{r}_{Qk} - \mathbf{r}_{Pj}$ with a negative strength $-f''_{Pj} f'_{Qk}$. Thus the only type of peaks that occur in the $\mathscr{P}_s(\mathbf{r})$ synthesis is the one corresponding to a vector from an anomalous (P) to a normal (Q) scatterer. It is also clear that these peaks can be described as consisting of P clusters of positive peaks, each cluster being an image of the Q atoms as seen from one P atom, and another set of P clusters of negative peaks, which are, respectively, at locations inverse to the corresponding positive peaks. In principle, therefore, it is possible to extract the structure of Q atoms from this synthesis by the use of image-seeking functions. The resultant structure obtained in this way is also in its absolute configuration, provided the necessary precautions have been taken in indexing the reflections. Figure 2 shows the $\mathscr{P}_s(\mathbf{r})$ synthesis in the case of α quartz, as obtained by Zachariasen[15] using the very accurate intensity data mentioned in Ch. 11. In this case, there are three types of vectors, Si–Si, Si–O, and O–O. Of these, the only vectors between unlike atoms are the Si–O vectors. It will be seen from Fig. 2 that these fall into two classes, one set corresponding to the positive peaks, and the other set, at inverse locations, corresponding to negative peaks, as predicted by theory.

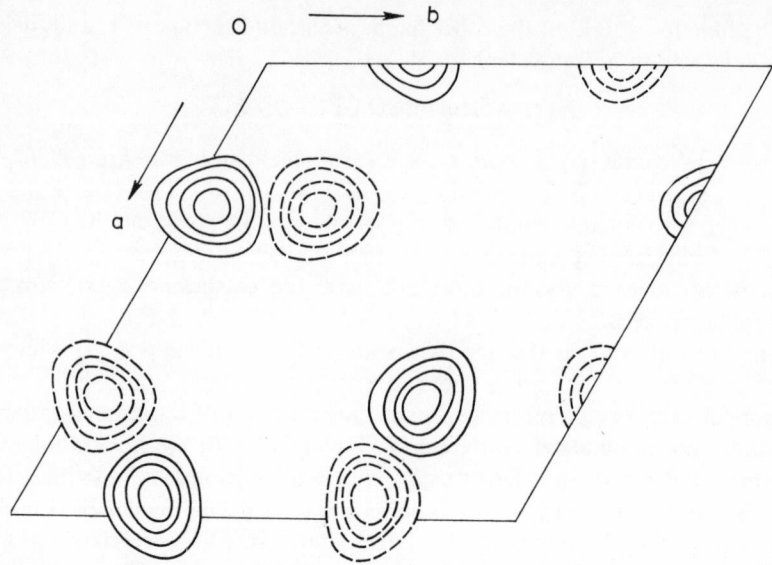

Fig. 2. The Patterson sine synthesis of α-quartz.

The function $\mathscr{P}_s(\mathbf{r})$ is still only a vector diagram. While the geometrical methods using image-seeking functions can be applied on it to extract the structure, the method becomes more difficult when the number of P atoms becomes larger. Even such a treatment of the function $\mathscr{P}_s(\mathbf{r})$ by image-seeking functions (say the weighted sum function) is only equivalent to the α_{an} synthesis. From a Fourier point of view, we have already seen that the β_{an} synthesis is more effective than the corresponding α synthesis. We shall also see later in Ch. 13 that the combination of the isomorphous-replacement and anomalous-dispersion methods through the Fourier approach, e.g., via the β_{is} and β_{an} syntheses, tends to be still more effective and a solution to the phase problem can be obtained better by this method than via the α type of syntheses.

Determination of the positions of anomalous scatterers

In order to be able to apply the anomalous-dispersion technique for the solutions of structures, it is obvious that the positions of the anomalously scattering atoms should first be known. Usually, the anomalous scatterers are also heavy atoms, so that the ordinary Patterson synthesis could be applied to determine their positions. However, situations may arise when such normal procedures do not lead to the location of the anomalous scatterers. In such cases, we would like to examine the possibilities of determining their positions by methods based on the Bijvoet differences, or some function of the observed differences in the intensities of inverse reflections.

Just as in the case of a pair of isomorphous crystals, we may examine first the method of the differences in intensity, $\Delta I \; [= I(\mathbf{H}) - I(\overline{\mathbf{H}})]$ as coefficients in a Patterson synthesis. However, we have already seen that when ΔI is used, we obtain the Patterson sine synthesis $\mathscr{P}_s(\mathbf{r})$. This function $\mathscr{P}_s(\mathbf{r})$ is devoid of any $[PP]$ vectors and is thus not useful, unlike the DP function met with in the isomorphous-replacement case. (This is true when all the P atoms are alike. Even when they are not alike the $\mathscr{P}_s(\mathbf{r})$ function has peaks whose magnitudes involve differences, such as $f'_{Pi} f''_{Pj} - f''_{Pi} f'_{Pj}$, and thus are not large.)

In this connection we may consider two other functions. One is the analog of the MD-squared synthesis, which, for the anomalous-dispersion case, is defined as the synthesis using the coefficients $(|\,|F(\mathbf{H})| - |F(\overline{\mathbf{H}})|\,|^2)$. Referring to Fig. 3 of Ch. 11, it is seen that, when $F''_P \ll F'_N$,

$$\Delta |F| = |F_N| - |\bar{F}_N| \simeq 2|F''_P| \cos \psi. \tag{25}$$

Thus

$$(\Delta |F|)^2 = 4|F''_P|^2 \cos^2 \psi$$

$$= 2(|F''_P|^2 + |F''_P|^2 \cos 2\psi)$$

$$= 2|F''_P|^2 + 2|F''_P|^2 \cos 2(\phi'_N - \phi''_P)$$

$$= 2|F''_P|^2 + |F''_P|^2$$

$$\times [\exp(2i\phi''_P) \exp(2i\phi'_N) + \exp(-2i\phi''_P) \exp(-2i\phi'_N)]. \tag{26}$$

We shall call this the MD_{an}-squared synthesis. This synthesis contains strong peaks corresponding to the Patterson of the P group arising from the first term of (26). The second term of (26) contributes only to the background, since it is a convolution of $|F''_P|^2$ with a quadruple convolution of the phase-squared and inverse phase-squared structures of F''_P and F'_N. This function $(\Delta |F|)^2$ has been used successfully by Rossmann[13] in protein crystallography. The MD_{an}-squared synthesis is more effective the smaller the mean contribution from F''_P as compared with F'_N and F_Q. This situation is parallel to the MD-squared synthesis in the isomorphous-replacement case. In both cases, if the two magnitudes whose differences are being taken are large compared with the vector being sought, then the difference in magnitude is a very good approximation to the magnitude of the vector difference. Fortunately, this is precisely the situation met with in complex structures such as those of proteins, where the heavy-atom contribution itself becomes a small fraction of the total intensity. Also, the anomalous contribution F''_P is, in general, small and hence this approximation is quite valid in the anomalous-dispersion case.

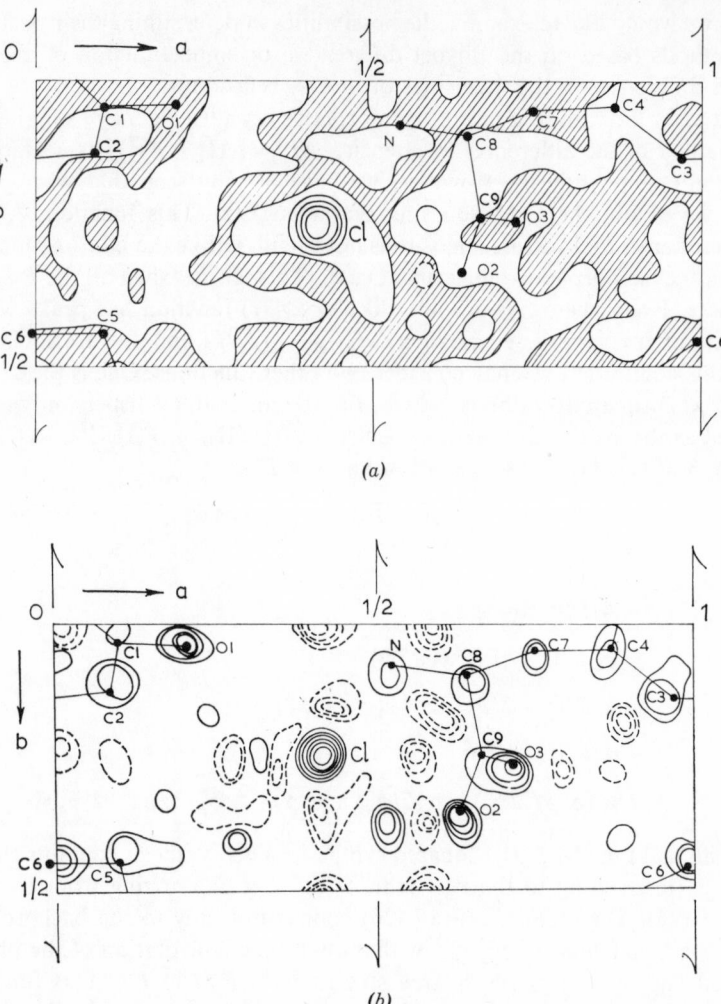

Fig. 3. The $\rho''(\mathbf{r})$ function calculated for the c projection of tyrosine hydro-chloride. (a) Using actual observed data of $F(\mathbf{H})$ and $F(\bar{\mathbf{H}})$ and calculated phases for $\mathbf{H}$ and $\bar{\mathbf{H}}$ including $\Delta f''$ for chlorine. (b) Using values of $F(\mathbf{H})$ and $F(\bar{\mathbf{H}})$ with phases including $\Delta f''$ for chlorine and also the light atoms.

In cases where the heavy-atom contribution (or the component F_P'') becomes appreciable, the approximation becomes invalid rapidly with increase in value of F_P''. An examination of the order of approximation involved can be made as follows. We have from (10) of Ch. 11,

$$\Delta I = 4\,|F_N'|\,|F_P''|\sin(\phi_N' - \phi_P'). \tag{27}$$

The MD_{an}-squared synthesis can therefore be written in the form

$$(|F| + |\bar{F}|) \, \Delta |F| = 4 |F'_N| \, |F''_P| \sin (\phi'_N - \phi'_P).$$

or

$$\Delta |F| = \frac{4 |F'_N| \, |F''_P|}{(|F| + |\bar{F}|)} \sin (\phi'_N - \phi'_P). \tag{28}$$

In (28) we could make the approximation $2|F'_N| \simeq |F| + |\bar{F}|$ so that

$$\Delta |F| \simeq 2 |F''_P| \sin (\phi'_N - \phi'_P). \tag{29}$$

Thus, the synthesis with $(\Delta |F|)^2$ as coefficients gives, as a first approximation, peaks corresponding to the Patterson $|F''_P|^2$, as may be seen readily by expansion of the right-hand side of $(\Delta |F|)^2$ using (30). Even the approximation involved in the above can be eliminated as follows. From (27), we have

$$\frac{\Delta I}{|F'_N|} = 2 |F''_P| \sin (\phi'_N - \phi'_P), \tag{30}$$

so that a synthesis with coefficients $(\Delta I)^2 / |F'_N|^2$ would lead to peaks corresponding to the Patterson of the P group, by the same argument as above. The quantity $|F'_N|^2$ can be obtained exactly by the relation (8) of Ch. 11, namely $|F'_N|^2 = \frac{1}{2}(|F|^2 + |\bar{F}|^2) - |F''_P|^2$. Since $|F''_P|^2 \ll |F'_N|^2$, this result also shows that, as a good first approximation, the synthesis with coefficients r^2, where r is the Bijvoet ratio [see (16), Ch. 11], namely

$$r = \frac{I(\mathbf{H}) - I(\bar{\mathbf{H}})}{\frac{1}{2}[I(\mathbf{H}) + I(\bar{\mathbf{H}})]}, \tag{31}$$

should also lead to the Patterson. This synthesis has an advantage that r is a dimensionless quantity and the errors of experiment are thereby minimized.[12]

A closely related function is $(\Delta I)^2$, suggested by Raman and Lipscomb.[12] Although this function is easier to interpret from the point of view of theory it is equivalent to giving much larger weights to strong reflections as compared to (31), or the MD_{an}-squared synthesis and would appear to be less preferable than these.

Fourier representation of anomalous-dispersion components[14]

In this section, we shall consider a method of Fourier representation of the dispersion and absorption components of the scattering material in the unit cell, corresponding to those which lead to the atomic scattering factors $f'_j = f^0_j + \delta f'_j$ and $f''_j = \Delta f''_j$. We shall take the general case in which all the anomalously scattering atoms in the P group are not necessarily alike. From (2a) and (2b), we have

$$F_N(\mathbf{H}) = F'_N(\mathbf{H}) + F''_P(\mathbf{H}), \tag{32a}$$

$$F_N(\overline{H}) = F'_N(\overline{H}) + F''_P(\overline{H}), \tag{32b}$$

where F'_N is the contribution to the structure factor from the dispersion components f'_{Pj} and f_{Qk} of the P and Q atoms, respectively. Since $\tilde{F}'_N(\overline{H}) = F'_N(H)$ and $\tilde{F}''_P(\overline{H}) = -F''_P(H)$, (32b) takes the form

$$\tilde{F}_N(\overline{H}) = F'_N(H) - F''_P(H). \tag{32c}$$

It follows from (32a) and (32c) that

$$\tfrac{1}{2}[F_N(H) + \tilde{F}_N(\overline{H})] = F'_N(H), \tag{33a}$$

so that it is the structure factor of the dispersion part of the scattering factors, and

$$\tfrac{1}{2}[F_N(H) - \tilde{F}_N(\overline{H})] = F''_P(H), \tag{33b}$$

which thus corresponds to the absorption part of the scattering factors. Thus we may write the electron-density distribution $\rho(\mathbf{r})$ of the crystal in the form

$$\rho(\mathbf{r}) = \frac{1}{V} \left\{ \sum_H \left[\frac{F_N(H) + \tilde{F}_N(\overline{H})}{2} \right] \exp(2\pi i H \cdot \mathbf{r}) \right.$$
$$\left. + \sum_H \left[\frac{F_N(H) - \tilde{F}_N(\overline{H})}{2} \right] \exp(2\pi i H \cdot \mathbf{r}). \right\} \tag{34}$$

It may be readily verified that the above expression reduces to the conventional one in the absence of anomalous dispersion, when the relation $F_N(H) \equiv \tilde{F}_N(\overline{H})$ holds.

From relation (33a), it is seen that the first summation on the right-hand side of (34) is equivalent to using the coefficients F'_N, and thus it will yield the structure corresponding to the dispersion components f'_N. We may refer to this as the structure $\rho'(\mathbf{r})$. On the other hand, (33b) yields

$$\frac{F_N(H) - \tilde{F}_N(\overline{H})}{2} = F''_P = i \sum_j \Delta f''_{Pj} \exp(2\pi i H \cdot \mathbf{r}_{Pj}),$$

so that the coefficients $-i[F_N(H) - \tilde{F}_N(\overline{H})]/2$ will lead to the structure with peaks of strength $\Delta f''_{Pj}$ at the positions of the anomalous scatterers $\mathbf{r}_{Pj}$, which corresponds to the absorption components of scattering factors. We may call this as the structure $\rho''(\mathbf{r})$. Thus $\rho(\mathbf{r})$, given by (34), can be represented as a complex function with real $[\rho'(\mathbf{r})]$ and imaginary $[\rho''(\mathbf{r})]$ components, in the form

$$\rho(\mathbf{r}) = \rho'(\mathbf{r}) + i\rho''(\mathbf{r}), \tag{35}$$

where

$$\rho'(\mathbf{r}) = \frac{1}{V} \sum_H \left[\frac{F_N(H) + \tilde{F}_N(\overline{H})}{2} \right] \exp(2\pi i H \cdot \mathbf{r}), \tag{36a}$$

and

$$\rho''(\mathbf{r}) = \frac{-i}{V} \sum_{\mathbf{H}} \left[\frac{F_N(\mathbf{H}) - \tilde{F}_N(\overline{\mathbf{H}})}{2} \right] \exp(2\pi i \mathbf{H} \cdot \mathbf{r}). \tag{36b}$$

We may obtain explicit expressions for $\rho'(\mathbf{r})$ and $\rho''(\mathbf{r})$ in terms of the real and imaginary components of the structure factors as follows. We have

$$A'(\mathbf{H}) = \sum_{j=1}^{Q} f_{Qj} \cos(2\pi\mathbf{H} \cdot \mathbf{r}_{Qj}) + \sum_{k=1}^{P} f'_{Pk} \cos(2\pi\mathbf{H} \cdot \mathbf{r}_{Pk}), \tag{37a}$$

$$A''(\mathbf{H}) = \sum_{k=1}^{P} \Delta f''_{Pk} \cos(2\pi\mathbf{H} \cdot \mathbf{r}_{Pk}), \tag{37b}$$

and exactly similar summations of sine functions for the B components. Thus

$$F(\mathbf{H}) = F'_N(\mathbf{H}) + F''_P(\mathbf{H})$$
$$= [A'(\mathbf{H}) - B''_P(\mathbf{H})] + i[B'(\mathbf{H}) + A''_P(\mathbf{H})], \tag{38a}$$

and similarly

$$F(\overline{\mathbf{H}}) = [A'(\mathbf{H}) + B''_P(\mathbf{H})] + i[A''_P(\mathbf{H}) - B'(\mathbf{H})], \tag{38b}$$

so that

$$\tilde{F}(\overline{\mathbf{H}}) = [A'(\mathbf{H}) + B''_P(\mathbf{H})] + i[B'(\mathbf{H}) - A''_P(\mathbf{H})]. \tag{38c}$$

Substituting these in (34), we obtain

$$\rho(\mathbf{r}) = \frac{1}{V} \{F(0) + 2 \sum{}' [A'(\mathbf{H}) + iB'(\mathbf{H})] \exp(2\pi i \mathbf{H} \cdot \mathbf{r})$$
$$+ 2i \sum{}' [A''_P(\mathbf{H}) + iB''_P(\mathbf{H})] \exp(2\pi i \mathbf{H} \cdot \mathbf{r})\}, \tag{39}$$

in which the prime over the summation symbol indicates that the sum is over one half of reciprocal space. Thus, from (35) and (39), we obtain

$$\rho'(\mathbf{r}) = \frac{1}{V} \{F(0) + 2 \sum{}' [A'(\mathbf{H}) + iB'(\mathbf{H})] \exp(2\pi i \mathbf{H} \cdot \mathbf{r})\} \tag{40a}$$

and

$$\rho''(\mathbf{r}) = \frac{2}{V} \sum{}' [A''_P(\mathbf{H}) + iB''_P(\mathbf{H})] \exp(2\pi i \mathbf{H} \cdot \mathbf{r}). \tag{40b}$$

These again show that the two functions $\rho'(\mathbf{r})$ and $\rho''(\mathbf{r})$ are the electron-density distributions corresponding to the dispersion (f'_N) and absorption ($\Delta f''_P$) components of the structure, and will contain peak strengths proportional to these at position $\mathbf{r}_{Nj}$ and $\mathbf{r}_{Pk}$, respectively.

The actual computation of these syntheses is readily made using expressions (36a) and (36b). We may note that the amplitudes to be used are $|F(\mathbf{H})|$ and $|\tilde{F}(\overline{\mathbf{H}})|$ ($\equiv |F(\overline{\mathbf{H}})|$), while their phases are deduced from (38a) and (38c) to be given by

$$\tan \phi(\mathbf{H}) = \frac{B'(\mathbf{H}) + A_P''(\mathbf{H})}{A'(\mathbf{H}) - B_P''(\mathbf{H})}. \tag{41a}$$

$$\tan \tilde{\phi}(\overline{\mathbf{H}}) = \frac{B'(\mathbf{H}) - A_P''(\mathbf{H})}{A'(\mathbf{H}) + B_P''(\mathbf{H})}. \tag{41b}$$

We also have

$$\tan \phi(\overline{\mathbf{H}}) = \frac{A_P''(\mathbf{H}) - B'(\mathbf{H})}{A'(\mathbf{H}) + B_P''(\mathbf{H})}. \tag{41c}$$

The synthesis $\rho''(\mathbf{r})$ can be used to determine $\Delta f_{Pk}''$ by successive approximations, provided accurate measurements of $|F(\mathbf{H})|$ and $|F(\overline{\mathbf{H}})|$ are available. Thus a Fourier map[14] of $\rho''(\mathbf{r})$ obtained, using measured data of Bijvoet differences for the c projection of L-tyrosine hydrochloride, is shown in Fig. 3a. Chlorine was the anomalous scatterer, whose dispersion and absorption components were used to calculate the phases in (41a) and (41b). It will be seen that clear peaks have developed at the positions of the chlorine atoms and very little background elsewhere. When the dispersion and absorption components of the light atoms were also included for calculating the phases, the resulting synthesis is shown in Fig. 3b. It will be noticed that peaks have appeared in this figure at the light-atom positions also in addition to those at the chlorine atoms.

Suppose, however, that, in calculating the synthesis, the correct values of $|F(\mathbf{H})|$ and $|F(\overline{\mathbf{H}})|$ are included, but the absorption component of the scattering factor is not taken into account in calculating the phases. Such a situation may arise, for instance, in a heavy-atom derivative of a protein, where measurable Bijvoet differences may be available, but the positions of the anomalous scatterers themselves may not be known to start with. We then have

$$\tan \phi_0(\mathbf{H}) = -\tan \phi_0(\overline{\mathbf{H}}) = \frac{B'(\mathbf{H})}{A'(\mathbf{H})} \tag{42}$$

and

$$\tan \phi_0(\mathbf{H}) \equiv \tan \tilde{\phi}_0(\overline{\mathbf{H}}), \tag{43}$$

so that (36b) reduces to

$$\rho''(\mathbf{r}) = -i \sum_{\mathbf{H}} [|F(\mathbf{H}) - |F(\overline{\mathbf{H}})||] \exp [i\phi_0(\mathbf{H})] \exp [2\pi i \mathbf{H} \cdot \mathbf{r}]. \tag{44}$$

It will be interesting to verify whether this synthesis exhibits peaks at the positions of the anomalous scatterers. This was tested[1] in the case of the above example of tyrosine hydrochloride and the resulting synthesis is shown in Fig. 4. It will be seen that some peak strength has developed in the chlorine positions, but it is much weaker than in Fig. 3a.

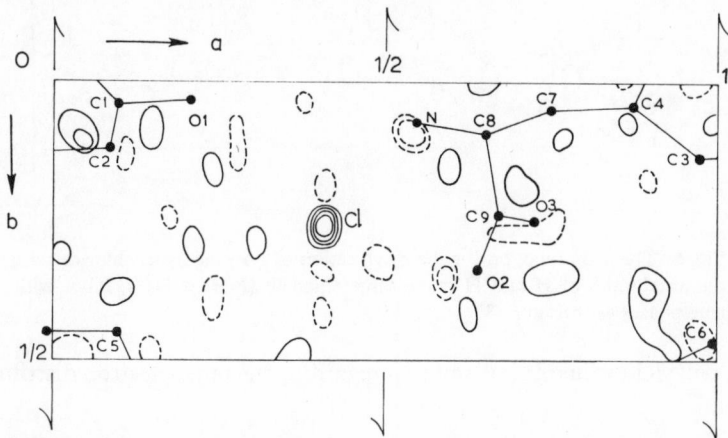

Fig. 4. The $\rho''(\mathbf{r})$ function for the c projection of tyrosine hydrochloride with true amplitudes as in Figure 3b but with phases same as for H without correction for $\Delta f''$.

It is of theoretical interest to consider what will happen if the correct values of the phases $\phi(\mathbf{H})$ and $\tilde{\phi}(\overline{\mathbf{H}})$, as given by (36a) and (36b), are used in a synthesis, but the magnitudes of $|F(\mathbf{H})|$ and $|F(\overline{\mathbf{H}})|$ are made equal. The resultant synthesis is shown in Fig. 5, in which there are quite strong peaks at the positions of the anomalous scatterers. This is to be expected, in analogy with the phase synthesis discussed in Ch. 4. However, as with the phase synthesis, this synthesis is only of theoretical interest, and cannot be used practically for determining either the positions of the P atoms, or their scattering factors $\Delta f_P''$. The few negative peaks in Figs. 3a and 5 have been shown to be series termination effects.[1]

It is obvious from (36a) and (40a) that the function $\rho'(\mathbf{r})$ includes in itself the component giving rise to the dispersion corrections $\delta f'_{P_k}$. Since this correction term is appreciable only when the incident wavelength is close to the absorption edge of anomalous scatterers, a difference-Fourier synthesis of the form $[\rho'_{\lambda_1}(\mathbf{r}) - \rho'_{\lambda_2}(\mathbf{r})]$ will give the electron-density distribution corresponding to $\delta f'$ at wavelength λ_1, provided λ_2 is chosen to be far removed from any absorption edges, so that $\delta f' = 0$ for λ_2 for all the atoms. The synthesis yielding $\rho''(\mathbf{r})$ and the difference-Fourier $[\rho'_{\lambda_1}(\mathbf{r}) - \rho'_{\lambda_2}(\mathbf{r})]$ mentioned

218 Anomalous dispersion effect: Fourier syntheses

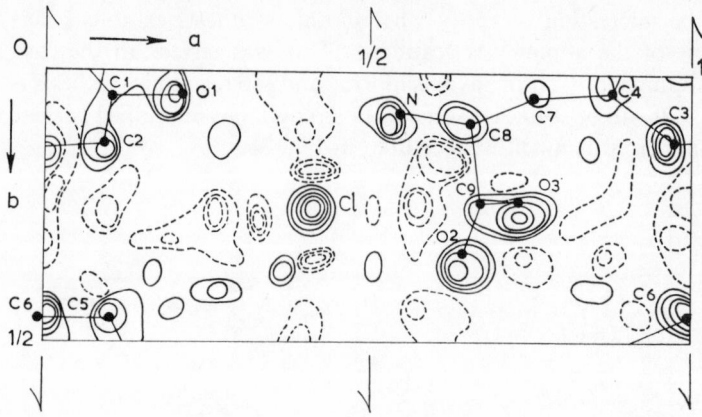

Fig. 5. The $\rho''(\mathbf{r})$ function for the c projection of tyrosine hydrochloride with the amplitudes of H and $\bar{H}$ made same equal to $[F(H) + F(\bar{H})]/2$ but with true phases as in Figure 3b.

above will yield valuable information regarding the inner electron distribution in atoms.

References

[1] K. K. Chacko and R. Srinivasan. *On the Fourier treatment of anomalous dispersion corrections in x-ray diffraction data.* Z. Kristallogr. **131** (1970) 88-94.

[2] G. Kartha. *Isomorphous replacement method in noncentrosymmetric structures.* Acta Crystallogr. **14** (1961) 680–686.

[3] V. Okaya and R. Pepinsky. *New-formulation and solution of the phase problem in x-ray analysis for noncentrosymmetric crystals containing anomalous scatterers.* Phys. Rev. **103** (1956) 1645–1657.

[4] Y. Okaya, Y. Saito, and R. Pepinsky. *New method in x-ray crystal structure determination involving the use of anomalous dispersion.* Phys. Rev. **98** (1955) 1858–1859.

[5] R. Pepinsky and Y. Okaya. *Determination of crystal structures by means of anomalously scattered x-rays.* Proc. Nat. Acad. Sci. U.S. **42** (1956) 286–292.

[6] R. Pepinsky and Y. Okaya. *Comparison of two procedures for solution of noncentric crystal structures utilizing anomalous dispersion.* Phys. Rev. **108** (1957) 1231–1232.

[7] R. Pepinsky, Y. Okaya, and Y. Takeuchi. *Theory and application of the $P_s(u)$ function and anomalous dispersion in direct determination of structures and absolute configurations in noncentric crystals.* Acta Crystallogr. **10** (1957) 756.

[8] G. N. Ramachandran and R. R. Ayyar. *Fourier syntheses for feeding in isomorphous and anomalous dispersion data.* In "Crystallography and crystal perfection" (Ed. G. N. Ramachandran) Academic Press, London. 1963. p. 25–41.

[9] G. N. Ramachandran and S. Raman. *Synthesis for the deconvolution of the Patterson function. Part I. General principles.* Acta Crystallogr. **12** (1959) 957–964.

[10] S. Raman. *Anomalous dispersion method of determining structure and absolute configuration of crystals.* Proc. Ind. Acad. Sci. **A47** (1958) 1–11.

[11] S. Raman. *Syntheses for the deconvolution of the Patterson function. Part II. Detailed theory for noncentrosymmetric crystals.* Acta Crystallogr. **12** (1959) 964–975.

[12] S. Raman and W. N. Lipscomb. *Two classes of functions for the location of heavy atoms and for solution of crystal structure.* Z. Kristallogr. **116** (1961) 314–327.

[13] M. G. Rossmann. *The position of anomalous scatterers in protein crystals.* Acta Crystallogr. **14** (1961) 383–388.

[14] R. Srinivasan and K. K. Chacko. *Fourier treatment of the anomalous dispersion corrections in x-ray diffraction data.* Curr. Sci. (India) **36** (1966) 279–281.

[15] W. H. Zachariasen. *Dispersion in quartz.* Acta Crystallogr. **18** (1965) 714–716.

13

Combination of isomorphous-replacement and anomalous-dispersion data: complete solution of phase

Choice of phase angle

From the results of the earlier chapters it is apparent that most of the properties associated with isomorphous-replacement and anomalous-dispersion methods are complementary. As mentioned there, this property enables us to compare the results of the two methods and then choose the result common to the two for arriving at the correct solution. This is evident, for instance, in the case of the phase ambiguity, for which the two possible values in the isomorphous-replacement case are of the type $\pm\theta$, while in the anomalous-dispersion case, they are θ and $\pi - \theta$ (when the atoms in the P group are all alike and have a centrosymmetric arrangement). The correct phase angle should therefore be the one common to both the cases. We have also developed two independent Fourier methods for arriving at the structure F_Q, namely that of the nonreplaceable, or the nonanomalous, scatterers, as the case may be, as well as syntheses for finding the locations of the P group of atoms; in these two methods the unwanted background had strengths of opposite sign. In this chapter, we shall develop suitable techniques of combining the two methods. As is to be expected, when the combination is properly made the phase problem has then a unique solution, and the unwanted data cancel out.

Before we consider the details of the formulas, the following comments are pertinent. When the replaceable atoms in the isomorphous crystals also exhibit anomalous dispersion, the structure factors in the various formulas corresponding to the P atoms are really F'_P, (arising from the total dispersion component of the P atoms) which includes also the dispersion correction $\delta f'_{Pj}$ in the atomic scattering factors. Thus in Fig. 1 of Ch. 9, the vectors F^I_P, F^{II}_P are really F'^I_P and F'^{II}_P, and hence $\delta F_P = F'^{II}_P - F'^I_P$. In order to avoid complication in notation, we may consider the following simplified example. (The results are readily extended to the general case in the light of the above discussion.)

Let $F'^I_P = 0$ in crystal I. This is equivalent to considering isomorphism of the addition type, so that the structure factor of crystal I may be denoted by F_Q. Let the crystal II, which has the P atoms in addition to those of crystal I, be denoted by structure F_N, so that we have the equations

$$F'_N = F_Q + F'_P, \tag{1}$$

and

$$F_N(\mathbf{H}) = F'_N + F''_P, \qquad F_N(\overline{\mathbf{H}}) = F'_N - F''_P, \tag{2}$$

where

$$F'_P = F'_P(\mathbf{H}), \qquad F''_P = F''_P(\mathbf{H}) \quad \text{and} \quad F_Q = F_Q(\mathbf{H}). \tag{3}$$

Thus the formulas used in the earlier chapters for both isomorphous-replacement and anomalous-dispersion cases separately can be considered, without appreciable alteration of the notation, for the combined case as well. Further, we shall concentrate our attention on the phase angle ϕ_Q of F_Q, which represents the "invariant" part of the structure. The quantities that are available from observations are $|F_Q|$, $|F_N(\mathbf{H})|$, and $|F_N(\overline{\mathbf{H}})|$. Knowing the positions of the P atoms, and their scattering factors f'_{Pj} and $\Delta f''_{Pj}$, F'_P and F''_P are known both in magnitude and phase. Hence, $|F'_N|$ can be obtained from the equation

$$|F'_N|^2 = \tfrac{1}{2}[|F_N(\mathbf{H})|^2 + |F_N(\overline{\mathbf{H}})|^2] - |F''_P|^2. \tag{4}$$

Combination of the beta types of syntheses

The β_{is} synthesis for the situation considered above is given by

$$\beta_{is} = \left[\frac{|F'_N|^2 - |F_Q|^2 - |F'_P|^2}{|F'_P|}\right] \exp(i\phi'_P), \tag{5}$$

which leads to

$$\beta_{is} \equiv F_Q + \tilde{F}_Q \exp(2i\phi'_P), \tag{6}$$

analogous to (18) of Ch. 10. The β_{an} synthesis is given by

$$\beta_{an} = \frac{\tfrac{1}{2}\Delta|F|^2 - (F'_P\tilde{F}''_P + \tilde{F}'_P F''_P)}{|F''_P|} \exp(i\phi''_P), \tag{7a}$$

which reduces to

$$\beta_{an} = \frac{\frac{1}{2}\Delta|F|^2}{|F_P''|}\exp\left[i\left(\phi_P' + \frac{\pi}{2}\right)\right],\tag{7b}$$

if the P atoms are all alike. In Ch. 12, this synthesis has been shown to be equivalent to

$$\beta_{an} = F_Q - \tilde{F}_Q\exp(2i\phi_P').\tag{8}$$

From (6) and (8) it follows that

$$\frac{\beta_{is} + \beta_{an}}{2} \equiv F_Q.\tag{9}$$

Hence if the coefficients given by (5) and (7) are added together and used in a Fourier synthesis, we would obtain the structure F_Q, with no additional background.

We showed earlier that the β_{is} and β_{an} syntheses are respectively equivalent to a double-phased synthesis using both the phases obtained in each case. The above combination of the two beta syntheses is thus equivalent to performing a Fourier synthesis with the unique phase ϕ_Q for F_Q, without, however, calculating it. On the other hand, it is obvious that a unique solution should exist for the phase angle ϕ_Q, Thus, using (3) of Ch. 9, we have

$$\cos(\phi_Q - \phi_P) = \frac{|F_N'|^2 - |F_Q|^2 - |F_P'|^2}{2|F_Q||F_P'|}.\tag{10}$$

The anomalous-dispersion case gives correspondingly a sine formula (when the P atoms are alike)

$$\sin(\phi_Q - \phi_P) = \frac{|F(\mathrm{H})|^2 - |F(\overline{\mathrm{H}})|^2}{4|F_Q||F_P''|},\tag{11}$$

so that

$$\tan(\phi_Q - \phi_P) = \tan\theta = \frac{\Delta I|F_P'|}{2|F_P''|(|F_N'|^2 - |F_Q|^2 - |F_P'|^2)}.\tag{12}$$

The Fourier synthesis for the Q atoms can therefore be performed using the observed $|F_Q|$ and the calculated phase $\phi_Q = (\phi_P + \theta)$, in which θ can be calculated using (12). Also one can obtain the calculated value of $|F_Q|$ for comparison with observed $|F_Q|$, using (10) and (11), which yield

$$|F_Q|\cos\theta = \frac{|F_N'|^2 - |F_Q|^2 - |F_P'|^2}{2|F_P'|} = X \text{ (say)}\tag{13a}$$

and

$$|F_Q|\sin\theta = \frac{\Delta I}{4|F_P''|} = Y \text{ (say)}\tag{13b}$$

so that

$$|F_Q|^2 = X^2 + Y^2. \tag{13c}$$

It was mentioned in Ch. 10 that, when anomalous dispersion effect is present, the variation of the $\delta f'$ correction with wavelength effectively affords an isomorphous-replacement-type of information. In fact, in principle, the phase problem can be solved uniquely by combining anomalous-dispersion data at two different wavelengths with a single crystal, without the need for two separate isomorphous crystals. This had been realized quite well by early workers.[10,12,14,16,17] For instance, in the case of a centrosymmetric crystal, the graphical method of sign determination discussed in Ch. 9 could be readily extended to the data with two wavelengths.[3,4,18] The noncentrosymmetric case has also been recently considered[20,21] and involves essentially formulas similar to the ones given above for the combination of isomorphous and anomalous dispersion data.

Location of heavy-atom positions using combination of data

The possible use of the isomorphous-replacement and anomalous-dispersion data individually for the location of the replaceable atoms, or the anomalous scatterers, as the case may be, was considered in Ch. 10 and 12. Each one, individually, could lead to a first approximation to the $[PP]$ vectors, e.g., the MD_{is}-squared synthesis and the MD_{an}-squared synthesis. We shall now show that a combination of the two types of data offers a more powerful method for the purpose than either of them separately.

We shall consider this aspect particularly with relevance to protein crystallography, where the case of isomorphism by addition of heavy atoms or groups is common. Besides the native protein and a series of heavy-atom derivatives, Bijvoet-difference data in certain cases may also be available. A practical aspect that has to be considered in this connection is that the series of isomorphous compounds may not show strict isomorphism. Tests are available, as considered in Ch. 10, for this purpose, which will be helpful in weighting the phase solution obtained by the usual methods. Assuming that isomorphism is good, we shall consider methods of determining the heavy-atom locations.

The formulas given in the last section involve quantities such as $|F_Q|$, or $|F_P''|$, in the denominator in (10) and (11). When these quantities are small, the coefficients tend to infinity. The effect may become particularly serious when there is lack of isomorphism, which will vitiate the values of $\sin(\phi_Q - \phi_P)$ and $\cos(\phi_Q - \phi_P)$ in (11) and (10). Moreover, in such complex structures as proteins, even the heavy-atom contribution $\langle|F_P'|^2\rangle$ forms only a small fraction ($\simeq 10\%$) of the mean total intensity and $|F_P''|^2$ would be still smaller. These circumstances warrant using an approximation in the formulas so as to make them involve quantities of the first power in the structure amplitudes. Thus for example, if we take the value of the Bijvoet difference ΔI from (11), we have

$$\Delta I = 4|F_Q||F_P''|\sin(\phi_Q - \phi_P'). \tag{14}$$

which can be written as

$$[|F_N(\mathbf{H})| - |F_N(\overline{\mathbf{H}})|][|F_N(\mathbf{H})| + |F_N(\overline{\mathbf{H}})|] = 4|F_Q||F_P''|\sin(\phi_Q - \phi_P'). \tag{15}$$

To a good degree of approximation, we can take $|F_N(\mathbf{H})| + |F_N(\overline{\mathbf{H}})| = 2|F_Q|$ in (15), so that

$$[|F_N(\mathbf{H})| - |F_N(\overline{\mathbf{H}})|] = 2|F_P''|\sin(\phi_Q - \phi_P'). \tag{16}$$

So also, in the case of isomorphous-replacement data, we get, from (10),

$$(|F_N'| - |F_Q|)(F_N'| + |F_Q|) = 2|F_Q||F_P'|\cos(\phi_Q - \phi_P') + |F_P'|^2, \tag{17}$$

or

$$(|F_N'| - |F_Q|) = \frac{2|F_Q||F_P'|}{(|F_N'| + |F_Q|)}\cos(\phi_Q - \phi_P') + \frac{|F_P'|^2}{(|F_N'| + |F_Q|)}. \tag{18}$$

Both $|F_N'|$ and $|F_Q|$ are large compared to $|F_P'|$. Thus the second term on the right-hand side of (18) can be neglected compared to the first. Further, making the approximation that $2|F_Q| \simeq (|F_N'| + |F_Q|)$, (18) reduces to

$$(|F_N'| - |F_Q|) = |F_P'|\cos(\phi_Q - \phi_P') \tag{19}$$

Thus we now have a pair of formulas (16) and (19), involving the product of the cosine and the sine of the angle $(\phi_Q - \phi_P')$ with $|F_P'|$ and $|F_P''|$, respectively, which gives the modulus difference in the two cases. Thus combining (16) and (19) we see that

$$\frac{1}{4k^2}[|F(\mathbf{H})| - |F(\overline{\mathbf{H}})|]^2 + [|F_N'| - |F_Q|]^2 = |F_P'|^2, \tag{20}$$

where

$$k = \frac{f_P''}{f_P'}, \tag{21}$$

The synthesis using the coefficients as given by the left-hand side of (20), which can be readily computed from the experimental data, thus lead purely to the Patterson of the P group. This may be compared with the individual MD-squared syntheses which yielded the Patterson of the P-group only approximately [(29), Ch. 10, and (26), Ch. 12]. It is readily verified that the background peaks contributed by the two individual terms on the left-hand side of (20) are equal, but of opposite strength, so that they cancel out completely in the sum, which just leads to the synthesis $|F_P'|^2$. The various treatments available in the literature[5-9,11] are based essentially on these linear formulas. These are valid, however, only to the extent of the validity of the approximation involved, which is fortunately applicable in protein crystallography.

We may now consider whether exact expressions could be obtained for the Patterson of the P group, similar to the phase solution discussed in the last section. We might start with the relations involving the actual intensities. Thus we have, from (10) and (11),

$$|F_P'| \cos (\phi_Q - \phi_P') = [|F_N'|^2 - |F_Q|^2 - |F_P'|^2]/2|F_Q| \tag{22a}$$

and

$$|F_P'| \sin (\phi_Q - \phi_P') = \Delta I/4k |F_Q|. \tag{22b}$$

Squaring and adding both sides of (22a) and (22b), we obtain a quadratic equation in $|F_P'|^2$, namely

$$|F_P'|^4 - 2|F_P'|^2[|F_N'|^2 + |F_Q|^2] + \frac{(\Delta I)^2}{4k^2} + (|F_N'|^2 - |F_Q|^2)^2 = 0. \tag{23}$$

If we denote by Δ and δ the difference in intensities corresponding to the anomalous dispersion and isomorphous cases, i.e.,

$$\Delta = \Delta I, \qquad \delta = (|F_N'|^2 - |F_Q|^2) = \delta I,$$

the solution of $|F_P'|^2$, from (23), is given by

$$|F_P'|^2 = (\delta + 2|F_Q|^2) \pm \left[(\delta + 2|F_Q|^2)^2 - \left(\frac{\Delta^2}{4k^2} + \delta^2 \right) \right]^{1/2}. \tag{24}$$

One of the two values of $|F_P'|^2$ deduced from (24) will be the correct one. This result is essentially the same as that of Singh and Ramaseshan,[19] who also consider the problem of obtaining the relative scale factor for the two data.

Correlation of heavy-atom positions in protein derivatives

The problem of correlating the relative positions of heavy atoms in a pair (or series) of isomorphous derivatives may also be tackled powerfully by using a combination of isomorphous-replacement and anomalous-dispersion data. The discussion below follows mainly the study made by Kartha and Parthasarathy,[6,7] but employs the notation and methods adopted in this monograph.

Let us consider for simplicity the linear formulas. Suppose, as in Ch. 10, that we have two heavy-atom derivatives I and II (structure factors F^{I} and F^{II}) of a parent protein, whose structure factor is F_Q. Denoting the heavy-atoms groups in the two crystals by P^{I} and P^{II}, we have the relations

$$F^{\mathrm{I}} = F'^{\mathrm{I}}_P + F_Q, \tag{25a}$$

$$F^{\mathrm{II}} = F'^{\mathrm{II}}_P + F_Q. \tag{25b}$$

Suppose that the atoms in the two P groups (P^{I} and P^{II}) occupy different positions in the unit cells, besides also possibly differing in their scattering factors. The atoms are also assumed to scatter anomalously, so that Bijvoet-difference data are also available for the two crystals I and II. We thus have the following quantities available from measurements on these two crystals, and on the parent protein:

$$[MD]^{\mathrm{I}}_{\mathrm{is}} = |F^{\mathrm{I}}| - |F_Q|, \qquad [MD]^{\mathrm{II}}_{\mathrm{is}} = |F^{\mathrm{II}}| - |F_Q|, \tag{26a}$$

$$[MD]^{\mathrm{I}}_{\mathrm{an}} = |F^{\mathrm{I}}(\mathbf{H})| - |F^{\mathrm{I}}(\overline{\mathbf{H}})|, \qquad [MD]^{\mathrm{II}}_{\mathrm{an}} = |F^{\mathrm{II}}(\mathbf{H})| - |F^{\mathrm{II}}(\overline{\mathbf{H}})|. \tag{26b}$$

From relation (19), we have

$$[MD]^{\mathrm{I}}_{\mathrm{is}} = |F'^{\mathrm{I}}_P| \cos(\phi_Q - \phi'^{\mathrm{I}}_P),$$

$$= \frac{|F'^{\mathrm{I}}_P|}{2} \{\exp[i(\phi_Q - \phi'^{\mathrm{I}}_P)] + \exp[-i(\phi_Q - \phi'^{\mathrm{I}}_P)]\}$$

$$= \tfrac{1}{2}[\tilde{F}'^{\mathrm{I}}_P \exp(i\phi_Q) + F'^{\mathrm{I}}_P \exp(-i\phi_Q)]. \tag{27a}$$

Similarly

$$[MD]^{\mathrm{II}}_{\mathrm{is}} = \tfrac{1}{2}[\tilde{F}'^{\mathrm{II}}_P \exp(i\phi_Q) + F'^{\mathrm{II}}_P \exp(-i\phi_Q)]. \tag{27b}$$

Also, from relation (16), we have

$$[MD]^{\mathrm{I}}_{\mathrm{an}} = 2|F''^{\mathrm{I}}_P| \sin(\phi_Q - \phi'^{\mathrm{I}}_P). \tag{28a}$$

Assuming all the P atoms in group I to be alike, and to have a value k^{I} for the constant k defined by (21), we have

$$[MD]^{\mathrm{I}}_{\mathrm{an}}/2k^{\mathrm{I}} = 2|F'^{\mathrm{I}}_P| \sin(\phi_Q - \phi'^{\mathrm{I}}_P). \tag{28b}$$

The right-hand side of (28b) can be expanded in a form similar to (27a), giving

$$\frac{[MD]^{\mathrm{I}}_{\mathrm{an}}}{2k^{\mathrm{I}}} = -\frac{i}{2}[\tilde{F}'^{\mathrm{I}}_P \exp(i\phi_Q) - F'^{\mathrm{I}}_P \exp(-i\phi_Q)]. \tag{29a}$$

Similarly,

$$\frac{[MD]^{\mathrm{II}}_{\mathrm{an}}}{2k^{\mathrm{II}}} = -\frac{i}{2}[\tilde{F}'^{\mathrm{II}}_P \exp(i\phi_Q) - F'^{\mathrm{II}}_P \exp(-i\phi_Q)]. \tag{29b}$$

Table 1 Comparison of correlation functions[7]

Coefficient	Origin peak	Correlation vectors $(\mathbf{r}_{Pi} - \mathbf{r}_{Pj}^{II})$	Centrosymmetric mate vectors $(\mathbf{r}_{Pj}^{II} - \mathbf{r}_{Pi}^{I})$	Self Patterson of the heavy atom $\pm(\mathbf{r}_{Pi}^{I} - \mathbf{r}_{Pj}^{I})$ and $\pm(\mathbf{r}_{Pi}^{II} - \mathbf{r}_{Pj}^{II})$	Back-ground	Symmetry
$[\|F^I\| - \|F^{II}\|]^2$	Present	Present with negative strength	Present with negative strength	Present	Present	Patterson
$[MD]_{is}^I, [MD]_{is}^{II}(= X^I X^{II})$	Absent	Present	Present	Absent	Present	Patterson
$[MD]_{an}^I [MD]_{an}^{II}/4k^I k^{II}$ $(= Y^I Y^{II})$	Absent	Present	Present	Absent	Present	Patterson
$[X^I X^{II} + Y^I Y^{II}]$	Absent	Present	Present	Absent	Absent	Patterson
$(X^I + iY^I)[X^{II} - iY^{II}]$	Absent	Absent	Present	Absent	Absent	Intensity symmetry, including anomalous dispersion
Origin correlation function (see text)	Only a single positive peak at $\mathbf{R}_{12}$					

If we denote the quantities on the left-hand sides of (27a), (27b), (29a), and (29b) by X^I, X^{II}, Y^I, and Y^{II}, respectively, then different combinations of the X's and the Y's can be shown to yield information about the vectors joining atoms in the P^I group with those in the P^{II} group. Thus, the expression $X^I X^{II} + Y^I Y^{II}$ can be shown to be equivalent to

$$X^I X^{II} + Y^I Y^{II} = \tfrac{1}{2}[F_P'^I \tilde{F}_P'^{II} + \tilde{F}_P'^I F_P'^{II}]. \tag{30}$$

A synthesis with these coefficients will thus have peaks at the set of positions $\mathbf{r}_{Pi}^I - \mathbf{r}_{Pj}^{II}$ and its centrosymmetric mate. On the other hand, if we consider the coefficients $(X^I + iY^I)(X^{II} - iY^{II})$, they are equivalent to $\tilde{F}_P'^I F_P'^{II}$, so that there will only be peaks at $\mathbf{r}_{Pi}^{II} - \mathbf{r}_{Pj}^I$. Several other combinations can be similarly examined, and the results are summarized in Table 1, following Kartha and Parthasarathy.[7] In certain cases, it might happen that the configuration of the P groups in both crystals are known, and the only unknown parameter is the vector $\mathbf{R}_{12}$ relating the origin of derivative II with respect to the origin of derivative I. Referring all coordinates to the origin of derivative I, we have

$$\mathbf{r}_{Pj}^{II} = \mathbf{R}_{12} + \mathbf{r}_{Pj\star}^{II}, \tag{31a}$$

$$F_P'^I = |F_P'^I| \exp(-i\phi_P'^I), \tag{31b}$$

$$F_P'^{II} = |F_P'^{II}| \exp(i\phi_{P\star}'^{II}) \exp(i\mathbf{R}_{12} \cdot \mathbf{H}) \tag{31c}$$

In these, $\mathbf{r}_{Pj\star}^{II}$ are the coordinates of the atoms P_j^{II} referred to the origin of crystal II and the corresponding phases are indicated by $\phi_{P\star}^{II}$. (The star is used to represent that the quantities refer to the origin of system II, and not system I.) Thus

$$\exp(i\mathbf{R}_{12} \cdot \mathbf{H}) = \frac{(X^I + iY^I)(X^{II} - iY^{II})}{|F_P'^I| \exp(-i\phi_P'^I)|F_P'^{II}| \exp(-i\phi_{P\star}'^{II})} \tag{32}$$

and the synthesis using the coefficient as given by the right-hand side of (32) will give a single strong peak corresponding to the vector $\mathbf{R}_{12}$, joining the origins of the two crystals I and II. This is termed the origin-correlation function.

Applications to neutron crystallography

The Fourier methods and other results discussed in this monograph are applicable to crystal structure determination by neutron diffraction also. Neutron diffraction has mostly played a complementary role to x-ray diffraction, as, for example, in the location of hydrogen atoms which are difficult to locate by x-ray diffraction.[1] Good reviews and standard books are available on this subject. However, there appears to be an interesting new possibility of applying neutron diffraction to large molecules, and this is concerned with the existence of anomalous scattering effects in neutron diffraction. In fact, this effect is quite pronounced in the neutron case and the correction terms arising from anomalous scattering could be as large as ten times the normal scattering component. The existence of this effect for neutrons was experimentally shown by Peterson and Smith [13] who performed the experiment analogous to the classical one of Coster, Knol, and Prins in the x-ray case mentioned in Ch. 11. The violation of Friedel's law was observed in a crystal of cadmium sulfide, with Cd as anomalous scatterer. In fact, when there is anomalous dispersion, the scattering length [†] can be represented by

$$b = b_0 + b' + ib'', \tag{33}$$

analogous to the x-ray case. Although theory predicted[2] a phase lag of $\pi/2$ for the absorption component, the experiments of Peterson and Smith showed, by comparison with the x-ray case, that this component is ahead of the dispersion component by $\pi/2$, just as in the case of x-rays. The possibility of applying these effects to the solution of the phase problem in crystallography, similar to the x-ray anomalous-dispersion technique, was pointed out by Peterson and Smith[13] and has been examined more closely by Ramaseshan.[15,20].

The main interest in the case of neutron anomalous-dispersion effect is that the correction terms b' and b'' are not small (unlike the x-ray case) and could be several orders of magnitude larger than the normal component b_0. Thus, for example, while f'/f_0 and f''/f_0 have maximum values of about 0.15 and 0.3 for the commonly used wavelengths of x-rays, the corresponding quantities in the neutron case could be as large as 5 and 10, respectively. Also, the behavior of the curves of b' and b'' as a function of frequency is quite different (Fig. 1).

A number of possibilities arise from these features. For instance, in the absence of the dispersion effect, the normal component b_0 itself does not exhibit large variations between different common elements, so that it is not possible to realize the situation corresponding to the "heavy atom" of the x-ray case. On the other hand, in the presence of anomalous dispersion with a sufficiently large value of b', it is possible to have effectively such a heavy-atom scattering. Further, with the proper choice of two wavelengths we could have effectively the data corresponding to both isomorphous-replacement, as well as anomalous-dispersion effects, so the discussions given earlier in this chapter concerning the possible unique solution of phase problem by the combination of isomorphous and anomalous data are equally valid for this case.[20,21] The main advantage is that, by using the two-wavelength method, isomorphism becomes exact.

[†] In the case of neutron diffraction, physicists usually use the term scattering length, which corresponds to the scattering factor in the x-ray case.

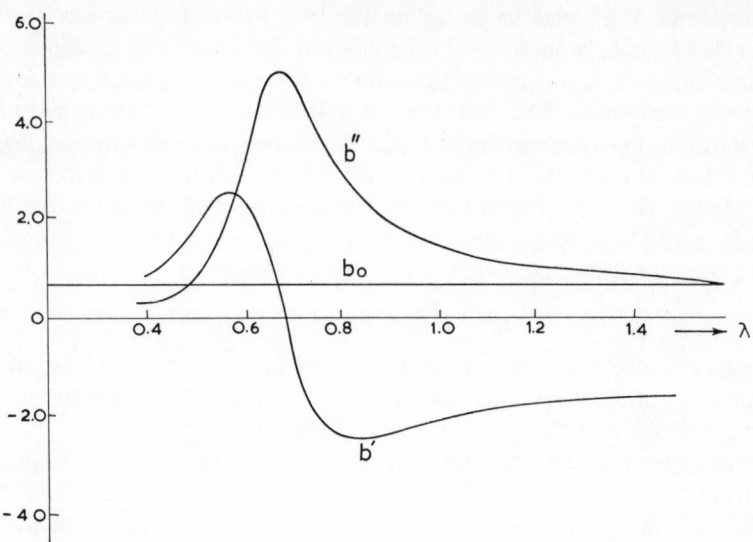

Fig. 1. Anomalous dispersion correction factors for neutrons. Variation of the normal (b_0), dispersion (b'), and absorption (b'') components as a function of wavelength.

Experimental verification

Some of these possibilities have been verified very recently. Two crystal structures were solved using such techniques by a young Indian from the Atomic Research Establishment, Bombay, who did the experiments at Harwell.[7a,17a,17b]

The first crystal whose structure was solved from anomalous dispersion data[7a] was that of cadmium nitrate tetradeuterate. In this structure the cadmium atoms, which exhibit anomalous dispersion, occur at special positions, and therefore their locations are known. However, the usual heavy-atom method could not be used to find out the rest of the structure, as the mean contribution (σ_1^2) of the cadmium atoms to the total intensity was only 1.4%. However, since the absorption component of cadmium (b''_{Cd}) is appreciable, the method of Ramachandran and Raman (Ref. 35, Ch. 11) was employed to calculate the phases and the double-phased synthesis [(24) of Ch. 11] was computed. As expected, positive peaks developed at most of the unknown atomic positions, as may be seen from Fig. 2. There were no peaks at the positions of the anomalously scattering cadmium atoms themselves, which is also to be expected from theory.

The sine Patterson function discussed in Ch. 12 was also tested in this case, and the resulting map was treated by using an 8-fold sum function, using the eight positions of the cadmium atoms. Although there were clear peaks, of different strengths at the expected atomic sites, there were "many other spurious peaks."

Incidentally, it was found that the values of phase angles as determined from anomalous effects in the pairs of inverse reflections agreed with the correct values to within about 30°. Table 2 lists the two ambiguous values

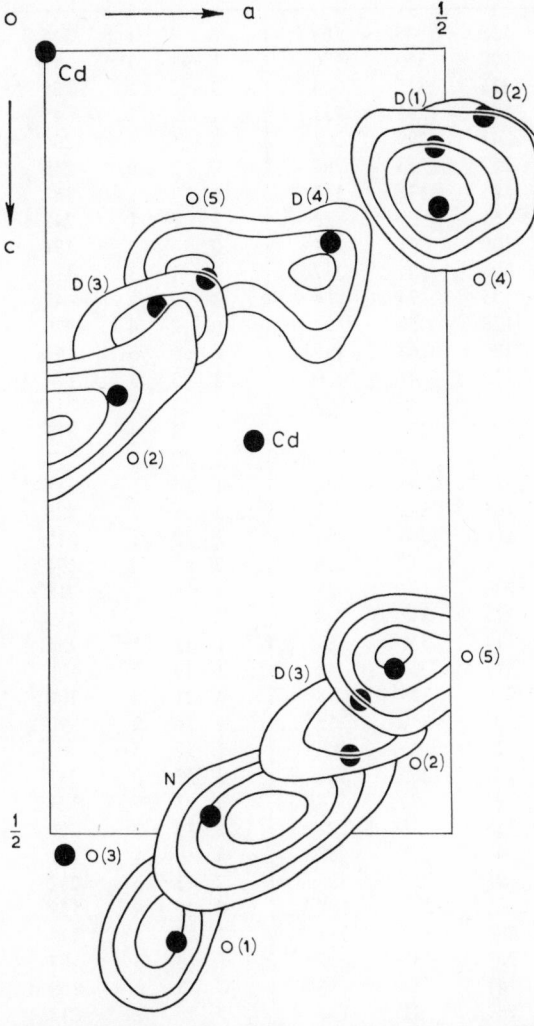

Fig. 2. Composite map of the double-phased anomalous Fourier synthesis. The contours are at equal but arbitrary intervals. The negative contours, as well as the zero and the first positive contour, are omitted. The correct atomic positions are indicated by black dots. Note that most of the atoms occur near the positive peaks.

Table 2   Values of the phase angles in cadmium nitrate tetradeuterate, determined from anomalous dispersion data, compared with those calculated from the structure finally determined[†7a]

h	k	l	ϕ_1	ϕ_2	ϕ_c	h	k	l	ϕ_1	ϕ_2	ϕ_c
0	2	2	135	45	109	3	1	1	315	315	18
1	1	1	100	350	76	1	11	3	94	356	148
1	3	1	315	315	304	3	3	1	165	285	300
1	5	1	160	291	271	1	13	1	45	45	18
0	6	2	241	299	203	2	6	4	206	334	330
1	1	3	257	13	286	0	2	6	249	291	304
1	7	1	135	135	117	1	7	5	182	88	110
1	3	3	61	29	22	3	5	1	248	22	250
0	0	4	109	71	77	0	14	2	191	349	190
1	5	3	306	324	337	3	1	3	116	334	354
2	0	2	155	24	24	3	7	1	45	45	103
0	4	4	128	52	120	0	12	4	230	310	326
1	9	1	186	264	190	0	6	6	157	23	145
2	4	2	139	41	0	3	3	3	174	96	106
1	7	3	45	45	36	1	13	3	236	34	26
0	10	2	270	270	290	1	9	5	176	274	328
1	11	1	206	64	63	2	12	2	173	7	137
1	9	3	217	53	102	1	15	1	315	315	321
0	8	4	198	342	1	3	5	3	225	225	202
1	1	5	184	266	194	3	9	1	315	315	286
2	8	2	134	47	34	2	10	4	194	346	158
1	3	5	233	37	25	3	7	3	216	54	209
2	2	4	228	312	288	2	0	6	139	41	117
1	5	5	114	336	63	1	11	5	236	34	26
1	15	3	113	337	41	1	19	3	88	1	264
0	10	6	216	323	295	3	11	5	108	342	311
1	3	7	102	348	255	4	10	2	90	90	71
3	9	3	225	225	188	1	17	5	169	281	336
1	17	1	281	269	162	3	15	3	315	315	233
1	5	7	190	80	78	3	3	7	174	95	191
3	1	5	135	135	125	1	13	7	190	80	238
3	3	5	45	45	58	3	17	1	221	49	218
1	13	5	91	359	335	3	13	5	202	68	70
1	7	7	225	225	195	0	20	4	137	43	112
4	2	2	234	308	237	2	18	4	174	6	317
2	8	6	203	337	0	1	5	9	97	352	305
0	16	4	141	38	86	0	12	8	178	2	61
0	18	2	159	21	134	2	20	2	218	322	255
3	11	3	279	351	54	5	3	1	215	55	83
0	0	8	124	56	86	0	22	2	135	45	46
2	16	2	199	341	358	3	17	3	173	277	111
1	17	3	237	33	78	1	15	7	45	45	3
2	14	4	239	301	212	0	18	6	230	310	232

Table 2 (*Continued*)

h	k	l	ϕ_1	ϕ_2	ϕ_c	h	k	l	ϕ_1	ϕ_2	ϕ_c
3	7	5	225	225	189	3	9	7	96	353	21
1	9	7	205	65	127	2	16	6	153	27	39
1	15	5	211	59	239	3	15	5	181	269	197
1	19	1	135	135	102	4	12	4	90	90	100
0	14	6	183	357	311	5	5	3	156	114	77
3	13	3	161	290	330	3	11	7	254	16	29
3	9	5	315	315	7	5	7	3	225	225	246
4	0	4	160	20	53	1	21	5	225	225	263
3	15	1	187	263	256	3	13	7	176	274	312
0	8	8	176	4	358	5	9	3	277	353	314
2	12	2	173	7	137	3	21	1	179	45	46
3	1	9	210	60	30	6	0	2	147	33	32
2	0	10	90	90	15	1	21	7	201	69	59
1	13	9	166	284	109	3	17	7	45	45	131
5	3	5	197	93	161	1	7	11	103	347	317
3	5	9	315	315	161	6	4	2	186	354	146
1	25	1	101	349	53	0	14	10	139	41	10
2	22	4	165	15	166	5	11	5	315	315	202
0	10	10	202	338	316	0	20	8	159	21	180
5	5	5	85	5	293	1	27	1	258	12	335
4	16	4	173	6	0	3	23	3	234	36	89
0	24	4	191	349	4	3	21	5	233	37	87
5	13	1	144	305	211	1	17	9	128	322	66
3	21	3	81	8	342	5	3	7	157	293	0
3	15	7	237	33	244	2	12	10	181	359	149
3	7	9	225	225	132	1	25	5	225	225	141
3	19	5	124	326	76	5	17	1	45	45	352
2	24	2	270	270	269	5	5	7	246	23	9
2	20	6	183	357	97	3	13	9	182	87	87
1	23	5	135	135	296	0	0	12	194	346	282
1	15	9	315	315	0	1	27	3	70	20	40
1	25	3	279	351	290	4	20	4	159	21	330
2	8	10	127	53	61	0	4	12	174	5	212
0	26	2	110	70	81	4	22	2	154	26	178
4	14	6	214	326	201	1	19	9	241	28	52
5	13	3	297	333	68	5	17	3	238	32	7
5	9	5	60	30	29						
3	23	1	140	310	307						
1	5	11	215	55	93						

† ϕ_1, ϕ_2 are the two values of the phase angles which are obtained from the measured Bijovet differences using (10) and (14). of Ch. 11. ϕ_c is calculated from the coordinates of the atoms in the structure. Note that one of ϕ_1 or ϕ_2 is close to ϕ_c in most cases.

obtained from the formula of Ramachandran and Raman (Ref. 35, Ch. 11), which are contained in (10) and (14) of Ch. 11. It is seen that, in every case, *one* of the calculated phases agrees reasonably well with the correct value.

In the case of the second crystal, namely samarium bromate nonahydrate which was centrosymmetric, the structure was solved by using the method of simulating isomorphism by the use of different wavelengths. The difference-Patterson synthesis, described in Ch. 10, which was applied in the case of x-rays by Ramaseshan, Venkatesan, and Mani (Ref. 40, Ch. 10), was employed with neutrons, using wavelengths of 1.27 and 0.88 Å, on either side of the resonance maximum of samarium, which is the anomalous scatterer in this crystal. This synthesis was then interpreted to locate the samarium atoms, which were at special positions. Knowing the locations of the samarium atoms, most of the other atoms in the asymmetric unit could be located by an analysis of interatomic vectors, and the complete structure was refined from this starting point. Sections of the anomalous scattering difference-Patterson map, together with its interpretation in terms of vectors, is shown in Figs. 3a and 3b.

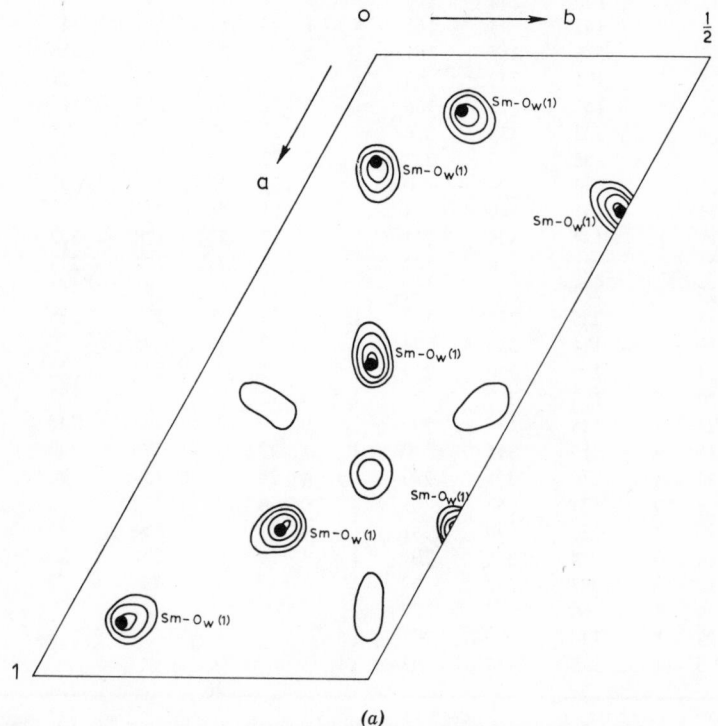

(a)

Fig. 3. The anomalous difference-Patterson map obtained from data at two different wavelengths of samarium bromate nonahydrate. (a) Section $z = \frac{1}{4}$;

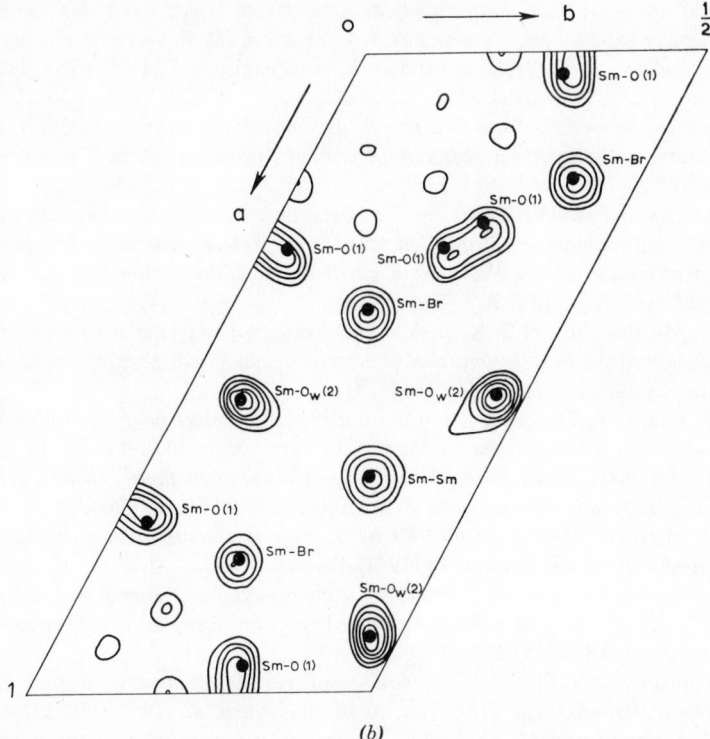

Fig. 3. (b) section $z = \frac{1}{2}$. The contours are at equal, but arbitrary, intervals. The negative, zero, and first positive contours are omitted. The relevant interatomic vectors are indicated in the diagrams.

Incidentally, it was found that the symbolic-addition procedure could be used for finding out the signs of structure factors with neutron-diffraction data in the case of centrosymmetric crystals, in spite of the fact that some of the atoms had negative atomic-scattering amplitudes.[17b] We shall not discuss this in detail, as the subject of direct methods is beyond the scope of this monograph.

References

[1] Bacon, G. E. *Neutron diffraction.* (Oxford University Press, London, 1962).

[2] G. Breit and E. Wigner. *Capture of slow neutrons.* Phys. Rev. **49** (1936) 519–531.

[3] S. Caticha-Ellis. *On the use of anomalous scattering of x-rays in the solution of centrosymmetric structures.* Acta Crystallogr. **15** (1962) 863–865.

[4] A. C. Hazell. *The application of the anomalous scattering of x-rays to the solution of the crystal structure of* $(NH_4)_2VO(NCS)_45H_2O$. Acta Crystallogr. **17** (1964) 1155–1159.

236   Isomorphous-replacement and anomalous-dispersion data

[5] G. Kartha. *Combination of multiple isomorphous replacement and anomalous dispersion data for protein structure determination. III. Refinement of heavy-atom positions by the least squares method.* Acta Crystallogr. **19** (1965) 883–885.

[6] G. Kartha and R. Parthasarathy. *Combination of multiple isomorphous replacement and anomalous dispersion data for protein structure determination. I. Determination of heavy-atom positions in protein derivatives.* Acta Crystallogr. **18** (1965) 745–749.

[7] G. Kartha and R. Parthasarathy. *Combination of multiple isomorphous replacement and anomalous dispersion data for protein structure determination. II. Correlation of the heavy-atom positions in different isomorphous proteins.* Acta Crystallogr. **18** (1965) 749–753.

[7a] A. C. Macdonald and S. K. Sikka. *The determination of the crystal structure of cadmium nitrate tetradeuterate using neutron anomalous dispersion measurements.* Acta Crystallogr. **B25** (1969) 1804–1811.

[8] B. W. Mathews. *The extension of the isomorphous replacement method to include anomalous scattering measurements.* Acta Crystallogr. **20** (1966) 82–86.

[9] B. W. Mathews. *The determination of the position of anomalously scattering heavy-atom groups in protein crystals.* Acta Crystallogr. **20** (1966) 230–239.

[10] C. M. Mitchell. *Phase determination by the two-wavelength method of Okaya and Pepinsky.* Acta Crystallogr. **10** (1957) 475–476.

[11] A. C. T. North. *The combination of isomorphous replacement and anomalous scattering data in phase determination of non-centrosymmetric reflections.* Acta Crystallogr. **18** (1965) 212–216.

[12] R. Pepinsky and Y. Okaya. *Determination of crystal structures by means of anomalously scattered x-rays.* Proc. Nat. Acad. Sci. Wash. **42** (1956) 286–292.

[13] S. W. Peterson and H. G. Smith. *Anomalous neutron diffraction in α-cadmiun sulfide.* Phys. Rev. Letters **6** (1961) 7–9.

[14] S. Raman. *Theory of the anomalous dispersion method of determining the structure and absolute configuration of noncentrosymmetric crystals.* Proc. Ind. Acad. Sci. **50A** (1959) 95–107.

[15] S. Ramaseshan. *The use of anomalous scattering of neutrons in the solution of crystal structures containing large molecules.* Curr. Sci. (India) **35** (1966) 87–91.

[16] S. Ramaseshan and K. Venkatesan. *The use of anomalous scattering without phase change in crystal structure analysis.* Curr. Sci. (India) **26** (1957) 352–353.

[17] S. Ramaseshan, K. Venkatesan, and N. V. Mani. *The use of anomalous scattering for the determination of crystal structures—KMnO₄.* Proc. Ind. Acad. Sci. **46A** (1957) 95–111.

[17a] S. K. Sikka. *The use of neutron resonance scattering in the structure determination of $Sm(BrO_3)_3 \cdot 9H_2O$.* Acta Crystallogr. **A25** (1969) 621–626.

[17b] S. K. Sikka. *On the application of the symbolic addition procedure in neutron diffraction structure determination.* Acta Crystallogr. **A25** (1969) 539–543.

[18] A. K. Singh. *The use of x-ray anomalous scattering in structure analysis of centrosymmetric crystals.* Proc. Ind. Acad. Sci. **66A**, (1968) 222–231.

[19] A. K. Singh and S. Ramaseshan. *The determination of heavy-atom positions in protein derivatives.* Acta Crystallogr. **21** (1966) 279–280.

[20] A. K. Singh and S. Ramaseshan. *The use of anomalous scattering in crystal structure analysis. I. Non-centrosymmetric structures.* Acta Crystallogr. **B24** (1968) 35–39.

[21] R. Srinivasan and K. K. Chacko. *On the determination of phases of a non-centro symmetric crystal by the anomalous dispersion method.* Z. Kristallogr. **131** (1970) 29–39.

Author Index

Number in bold denotes the appearance of the name in the reference list

239

Subject Index

Absolute configuration, 168, 191, 209
  and double phased anomalous
    dispersion synthesis, 191
  determination by anomalous
    scattering, 168
  determination from $P_s(\mathbf{r})$ function, 209
Absorption component, determination of,
    193
  inaccuracy in coordinates due to
    neglect of, 196
  need for inclusion in refinement, 196
  of atomic scattering factor, 167
Absorption edge, and atomic scattering
    factor, 7
Alpha anomalous synthesis, 202, 203, 210
  and $P_s(\mathbf{r})$ function, 210
  comparison with α_{is} synthesis, 202
  negative duplication of structure in,
    203
  peaks in, 202, 203
Alpha difference synthesis, peaks in, 78
Alpha general synthesis, 73–76, 82, 132
  background peaks in, 75
  comparison with α_{is} synthesis, 132
  comparison with β_{gen} synthesis, 82
  deduction of peaks in, 74
  for a centrosymmetric crystal, 76
  known peaks in, 74
  normalised peak ratio in, 75
  normalised peak strengths in, 75
  peaks in (centrosymmetric case), 76
  peaks in (noncentrosymmetric case), 74
  strength of known peaks in, 75
  strength of unwanted peaks in, 75
  strength of wanted peaks in, 75
  unwanted peaks in, 74
  wanted peaks in, 74
Alpha isomorphous synthesis, 132, 133
  background peaks in, 133

Alpha isomorphous synthesis (*continued*)
  coefficients for, 132
  comparison with α synthesis, 132
  suppression of replaceableatom peaks
    in, 133
Alpha modified synthesis, 86
Alpha prime modified synthesis, 87
Alpha prime synthesis, 80, 81, 85
Alpha synthesis, 54, 73, 77, 78, 80, 81,
    92, 96, 97, 109, 110, 111, 131, 201
  as convolution of $|F|^2$ and $|F_P|$, 73
  as weighted, modified sum function, 54
  as weighted sum function, 73, 96, 97
  comparison with β synthesis, 109, 110
  comparison with geometrical method,
    110
  comparison with sum function, 111
  'difference' type of, 78
  effect of wrong atoms in, 77
  for isomorphous replacement, 131
  general form of, 73
  interpretation of, 73
  modified form of, 77
  peak strength at input atoms in, 78
  peak strengths in, 92
  peaks in, 73
  peaks in, with wrong atoms, 77
  reduced peak strength at wrong atoms
    in, 78
  test of, 97, 111
  with anomalous dispersion data, 201
  with wrong atoms, 94
Ambiguity, in phase (anomalous
    dispersion case), 17
  in phase (isomorphous case), 126
  in sign of phase angle, with
    isomorphous data, 123
Analytical approach, to superposition
    function, 53

242

Subject index

DATE DUE

GAYLORD